PLAN
OF
YORK HARBOUR, &c.

Scale.

LOWER

MIDLAND

JOHNSTOWN

DISTRICT

CASTLE

DISTRICT

ISTRICT

CAN

EASTERN DIST

STATE

LAKE
CHAMPLAIN

VERMONT

LAKE ONTARIO

N   E   W   Y   O   R   K

OTTAWA DISTRICT

LOWER

DISTRICT

of   the

Middle Course of the Grand Canal

CANADA

Course   of   the   Grand   Canal

CONNECTICUT MASSACHUSETTS

NEW
JERSEY

NEW
YORK

LONG I⁰

YORK

CANADA

ATLANTIC OCEAN

Row, Dec.1.1820.   Neale & Son G. 352. Strand.

To Elwed Mac Millan

With the author's compliments

Lois Darroch Milani

May 2, 1974

Robert Gourlay, Gadfly

Robert Gourlay, Gadfly

THE BIOGRAPHY OF ROBERT (FLEMING) GOURLAY, 1778-1863
FORERUNNER OF THE REBELLION IN UPPER CANADA, 1837

by

LOIS DARROCH MILANI

AMPERSAND PRESS

PRINTED AND BOUND IN CANADA BY
JOHN DEYELL LIMITED

The first question in political economy
should be
can the mass of the people
live comfortably
under this or that
arrangement?
but this most necessary question
was forgotten,
and many of the people have perished.

Robert Gourlay
Statistical Account of Upper Canada,
1822, p. ccccvii

The "family compact of Upper Canada
is composed of those members
of its society who,
either by their abilities and character,
have been honoured by the confidence
of the executive government,
or who, by their industry
and intelligence,
have amassed wealth.
The party, I own, is comparatively
a small one; but
to put the multitude
at the top
and the few
at the bottom
is a radical reversion
of that pyramid of society
which every reflecting man
must foresee
can end only by its downfall.

Sir Francis Bond Head,
A Narrative of the Canadian Rebellion,
1839, p. 465

What we desire
for ourselves
we desire
for all.

J. S. Woodsworth

Acknowledgments

IN SCOTLAND

I wish to thank the late Brigadier William Edmonstone Duncan, my first contact with Robert Gourlay's descendants; the late Mrs. Dorothy Duncan, wife of a great grandson, the late Commander John Duncan, and their daughter, Dr. Ursula Duncan, of Parkhill, Arbroath, who allowed me to use family papers and pictures; the late Rev. Edward Millar of Lynwood, Banchory, and his late wife Margaret, great-grandaughter of Robert Gourlay, for the Wilkie drawings; Mr. and Mrs. John Playfair of Baltilly, Ceres, Fife; Mr. and Mrs. Andrew Duncan who rescued Craigrothie House from disuse; and the National Library, Edinburgh.

IN ENGLAND

I wish to thank Lt. Col. and Mrs. F. Norman Jeans of the Manor House, Fisherton de la Mere and Deptford Farm, Wiltshire; their son, Lt. Col. Harold E. Cassels, who restored the house on Deptford Farm, Wylye, to the graciousness of Gourlay's time; Salisbury City Library and Museum; the British Museum; and Colonial Office Records.

IN CANADA

I am indebted to the late William R. Riddell, who as a lawyer disapproved Robert Gourlay but as a biographer was fascinated by him; Dr. J. M. S. Careless of the University of Toronto for his graduate course in The History of Old Ontario for which I read in 1957-8; the Canada Council for a $700 grant in 1962 for a return research trip to Britain; the Dominion and Ontario Archives; Miss Edith Firth, Associate Head, Metropolitan Toronto Central Library, and the staff of the Baldwin Room; the library of the University of Toronto; my late parents, William F. Darroch, who financed my university education during the depression years of 1929-1933, and Leonora Wolverton Darroch, who always encouraged me; my husband, Camillo D. Milani, who financed me during the time of writing and research; my daughter, Mrs. John (Cristine) Bayly, who assisted me with typing; and others whom space prevents my mentioning individually.

vii

To
ALONZO WOLVERTON
Wolverton, Ontario

1841-1925

Contents

ix

Illustrations

Preface

This is the first complete biography of Robert (Fleming) Gourlay who was banished from Upper Canada in 1819 on pain of death if he returned. The book was prompted by a desire to research the early life of a man who left Britain in 1817 at the age of thirty-nine to come to Upper Canada, who quickly achieved widespread political popularity, and then was thrown out of the province "like a spoilt jelly".

Knowledge of Robert Gourlay's career before he came to Canada was incomplete. He moved from his home in Scotland in 1809 with the sheriff at his heels because he advocated electoral reform, and left England in 1817 partly because of opposition to his efforts on behalf of the indigent agricultural labourers in the southern shires. The circumstances surrounding these brief facts were none too clear.

The course of Robert Gourlay's meteoric two-year career in Upper Canada was better known because he made too big a flash to be forgotten. He arrived at the beginning of the struggle for power between the elected and appointed bodies of the government that would culminate in the rebellion of 1837 and Lord Durham's *Report on the Affairs of British North America*. He did not remain on the periphery. Twice he was arrested and tried on a charge of seditious libel and twice he was acquitted. Again he was arrested, this time on the charge of failing to leave the province when ordered by two members of the legislative council. He refused to submit to what he considered was an arbitrary order, demanded a proper trial, and was jailed for six months while he waited for it. At his third trial, too ill to defend himself, he was found guilty and sentenced to banishment. Research disclosed that the solicitor general of the province removed the minute book of the assizes of the Midland District that recorded Gourlay's acquittal after his two trials there; and in addition, the son of the chief justice, who was acting as clerk of the court during the third trial, failed to include in the minutes of the assizes of the Niagara District the trial by which he was banished. Why did they do

xi

this, if they were proud of ridding the country of a dangerous radical?

On his return to Britain Gourlay published a two-volume *Statistical Account of Upper Canada* for which he was both commended and criticized. To draw attention to the plight of the poor he cracked flints on the roads of Wiltshire with the paupers. To emphasize the need for an investigation into his arbitrary banishment from Upper Canada, he "whipped" Lord Brougham in the lobby of the House of Commons. He wrote a sane book while imprisoned in London on a charge of insanity. Later when William Lyon Mackenzie solicited his aid during the Rebellion of 1837, he refused to help because he deplored violent methods—refused, even though earlier he had been considered a sufficient threat to the peace of the province that he had been banished by its government. Such an astonishing concatenation of events, the clash of personalities that it involved, and conflicting reactions to the man who has been called brilliant and crack-brained, martyr and poseur, dedicated reformer and inveterate scribbler, warranted research to throw light on the portions of Robert Gourlay's life that till now have been unknown.

This biography was made possible because of the record left by a man who believed in the rightness of his purpose so intensely that he registered every action in order to save his name from eternal misrepresentation. The record is sometimes repetitious, contradictory and egotistical, but never tepid. Because his life is described as often as possible in his own words, this biography may be accused of being biased. The style and selection of material were determined partially by the fact that there was sufficient for a narrative made vivid by dramatic details so often lost to history. For this reason also the biography is more a chronicle than an analysis, for it was considered better to present than to analyze the career of a man who lived life with flair, unmindful of the cost. The latter characteristic is rare in a canny dominion prone to belittle the efforts of certain vigorous personalities who assisted in its building, particularly those who travelled to the left of the middle road that it has been Canada's pride to follow.

Robert Gourlay was banished because he travelled to the left in the course of the nineteenth century movement in which the emerging force was called democracy or political equality and in which the ruling clique was termed the Family Compact in Upper Canada and the aristocracy in Britain. In the twentieth century the emerging force is named socialism or economic equality, and the ruling clique is known sometimes as The Establishment, sometimes as Big Business, and sometimes as Technocracy. In the interim great improvements have been made in plumbing and locomotion and in the manner in which political leaders address each other. The dichotomy of right and left still remains as indigenous a force in politics as sex in the psychology of Freud.

I

Youth

Robert Gourlay was born in 1778 in Scotland on his father's estate just outside the village of Craigrothie in the shire of Fife.

The Gourlays could trace their land and ancestry to the twelfth century. After King William the Lion had been sent to France in chains for daring to invade England, he was freed on condition that he acknowledge vassalage to Henry II. When William returned to Scotland in 1175, it is reported in the Book of Heralds, he was accompanied by Ingelramus de Gourlay (variously spelled Ingelrame de Gourlaye). In return for watching and warding against invaders, Ingelramus was granted some of the oldest held land in Fife at Kincraig, the promontory that guards the entrance to the Firth of Forth, and at Craigrothie about twelve miles inland.

The Gourlay family was proud to number among its ancestors also a man who had been martyred in the struggle for religious freedom from the papacy. In 1515 Norman Gourlay swore his priestly vows at St. Andrews. After he met Martin Luther in Germany, Norman Gourlay married and refused to leave his wife when commanded by Catholic Cardinal Beaton. Beaton ruled Fife from the great red sandstone cathedral and his castle on a crag beside the sea at St. Andrews. If Norman Gourlay had entertained himself with a hundred concubines, he would not have been chastised, but the Church forbade a priest to cleave to one true wife. Beaton, who had several illegitimate children himself, determined to rid himself of the heretic who defied his order. In August 1534, "Normound" Gourlay was secretly assassinated, but Beaton feared an uprising if his body were burned in Fife. It was taken to Greynseyd Rood on a road leading from Edinburgh to Leith. There it was burned on a pyre so high that the leaping flames might be seen across the firth as a lesson to the rebellious Fifers there.[1]

---

[1] R. Gourlay, *The Banished Briton and Neptunian, being a record of the life, writings, principles and projects of Robert Fleming Gourlay, Esq., now Robert Fleming Gourlay,* #15, p. 133; Gourlay Papers, OA., 1852, 1862.

This was not the end of religious altercation. Three-quarters of a century later, in St. Andrews in 1599, John Knox's followers, "Enraged at idols, mass and beads, Dang the cathedral doun!" The Gourlays were Presbyterian.

The Gourlay tradition in the political sphere was milder. After the parliaments of England and Scotland were united, John Gourlay of Craigrothie refused to join the Earl of Mar when he marched out of Fife in 1715 to support the Stuarts against the Hanovers. He was expedient enough to realize that peace under one ruler would bring prosperity to Britain. The Gourlays prospered, and in the last quarter of the eighteenth century a Gourlay named Oliver (John's grandson) was busy bringing the family fortunes to their climax.

Oliver Gourlay had served his apprenticeship to a writer (barrister) in Cupar, the capital of Fife, before he moved to Edinburgh where he mingled with its best business and intellectual life. When his father died, he returned to the family estate at Craigrothie to take his place as a leading country gentleman in the shire. First he bought out his brother, a bachelor minister who had no inclination to burden himself with the care of land with its lingering feudal obligations, tithe collections and rents. Then he purchased the farm of Scotstarvit that adjoined Craigrothie to the west, pledging himself to pay an annual feu duty of £50 to Scot. The next year he sold a third of it to Colonel Wemyss of Wemyss Hall, his neighbour to the north-east, for as much as he had paid for it all. The following year he purchased Newton and Broadleys near Auchtermuchty. He soon sold the part that adjoined the village for the price of the whole. In 1773 he married Janet Fleming, heiress of Baltilly , the land with a mill that lay between the villages of Craigrothie and Ceres. All this made him owner of a continuous tract of land extending from Ceres burn to the parish of Cults.

A golden age of peace and prosperity seemed to stretch ahead for Scotland and the Gourlays. The days of the witch hunts and fierce religious and border feuds were gone, scrapping over the ruler had ended in 1746 with the defeat of Bonnie Prince Charlie, and Scottish farmers were prospering as they adopted Turnip Townshend's new fodder for cattle and enlarged their knowledge of manures and crop rotation. No longer could agriculture be simply the dropping of seeds on scratched ground, nor could the cattle be allowed to become so weak that in the spring they would have to be carried to the new grass. Agriculture was the major industry on an island with a growing population clamouring to be fed. Landowners who could not adopt new methods were selling the land as worn out. Those with energy like Oliver Gourlay were snapping it up, eager to improve its cultivation. When he was at the height of his vigour, he supervised five farms from Craigrothie. He reorganized the fields, planted trees on the tree-starved hills, and promoted the building of roads and bridges so that farm produce could be marketed in the towns. In the shire of Fife he was known as a leader among the progressive lairds.

The land that Oliver Gourlay acquired put into his hands five votes for members to be sent to the British parliament in London.[2] In the hodge-podge eighteenth century British parliament, elected by pocket and rotten boroughs and parchment barons, Scotland possessed 3,253 votes for its million and a half population. Of these, 236 votes belonged to Fife, the largest number of any Scottish shire. Fife votes elected eleven members of parliament, nearly one-quarter of the forty-five members Scotland was allotted in the British parliament at the time of the union of 1707. Oliver Gourlay's five votes gave him considerable prominence in the shire, enough to retain the name of Whig at a time when the Whig party did not enjoy the confidence of the king. The Whigs were more likely than Tories to espouse the liberal principles that had helped to bring about two revolutions, the American and the French.

Though Britain was technically a democracy because it held elections for a parliament that met fairly regularly, its king was still a personal ruler. The king was George III who said, "I will have no reform in my time." His advisers held office only as long as they retained his confidence. When George came to the throne in 1760, the Whigs had been in power since the ousting of the Stuarts. When the Whigs failed to win victories in the Seven Years' War, George turned to the Tories whose monopoly would be even more lengthy. George III's favorite adviser now was William Pitt the Younger who had entered parliament as a Chathamite Whig.

Influenced momentarily by the liberal ideas of Locke and Rousseau and his Whig colleagues, Charles Grey and Charles James Fox, Pitt made tentative proposals for parliamentary reform, one of the pressing needs of the country. The king was horrified at the proposed changes. Pitt and others of the King's advisers who were turning farther right gradually metamorphosed into the early nineteenth century Tory party. Charles Grey remained Whig. He founded the Society of the Friends of the People in order to popularize liberal ideas among the intelligentsia. An English shoemaker, Thomas Hardy, founded the London Corresponding Society so that he could compare progress with like-minded people in other parts of the kingdom. Lower in the social order than Grey, he would soon be prosecuted for sedition and threatened with death. A Scottish lawyer named Thomas Muir, who organized two conventions of the Friends of the People in a few months, would be brought to trial as well. It was dangerous to be a Whig in England at the time of the revolution in France. It was doubly dangerous in Scotland, for any one

---

[2]*Niagara Spectator*, Aug. 6, 1818.

who espoused principles different from royal and Tory ideas could be accused of plotting against the right of the Hanover to rule.

Oliver Gourlay was not afraid to praise France and detest the illiberal policy of Pitt. Even when the Bourbons were deposed on August 10, 1792, and Britons who criticized their own government could be accused of treason, he saw no reason to curry favour by renouncing his Whig background. His place in the shire of Fife was secure and both parties came to court him for his five votes for parliament. He had few ambitions beyond Fife's boundaries. He would not turn his back on friends like Sir James MacIntosh who had written *Vindiciae Gallicae*, 1791, as his reasoned answer to Edmund Burke's attack on the "swinish multitude" in *Reflections on the Revolution in France and on the proceedings in certain societies in London relative to that event*, 1790.

Though revolution was convulsing France and though it was dangerous in Britain to debate politics, Oliver Gourlay's personal life was serene. Janet Fleming had given him two daughters. On the morning of March 24, 1778, he rode off to superintend his men who were planting trees on Scotstarvit hill behind the house at Craigrothie. As he gallopped back, a serving woman came running towards him with the news: his firstborn son had just been ushered into the light. The child was called Robert after his paternal grandfather. Robert Gourlay grew up at Craigrothie with the expectation of administering a considerable estate when his father should cease to be active. He must also prepare himself to take a responsible part in administering county affairs, for at this time local matters were supervised by the lairds as a free service to the community.

Robert's formal education began in the village school at Craigrothie. Because the parish school was two miles away, Oliver had donated the land for one nearer home and the building was erected by public subscription. Here Robert, with a little urging from the ferrule of Dominie Angus, was educated till he was twelve.[3] He was then sent to St. Andrews for three years' preparation for its university. There he read about Norman Gourlay in John Knox's *History of the Reformation,* and often paused curiously in the school hall in front of the cracked portrait of wicked Cardinal Beaton who had been responsible for the martyrdom of an ancestor whom he revered. At other times he would peer down the stony neck of the bottle dungeon where persecutor Beaton himself was finally thrown, and chase through the ruins of the cathedral razed by Knox.

Religion still pervaded Scottish life though physical violence had been replaced by dialectic. St. Andrews trained students for the ministry. The current controversy in the church of Scotland was whether a man's

---

[3]J. C. Dent, *The Story of the Upper Canadian Rebellion*, (Toronto, 1885) I, p. 35.

soul would be saved by predestination or free will. Discussion of the new concept of free will caused much soul-searching among the students. Thomas Chalmers, who had come from Anstruther to study for the ministry, was beginning to reject the strict Calvinism of his home teaching, much to the sorrow of his parents. Robert did not face such a struggle when he rejected predestination. The concept was foreign to his naturally questioning mind, his independent nature, and the example set him by liberal parents who placed right action above ritual or strict theology.

The issue that puzzled Robert during his last years at St. Andrews was the even newer concept of the perfectibility of man, not in sole relation to his God but with regard to his fellow man. It was incomprehensible that a perfect God had created a world imperfect enough to harbour poverty. It was atheist William Godwin who introduced Britain to the new word, "perfectibility", that he had taken from his reading of Rousseau and Diderot. Godwin claimed that society as well as man could achieve perfection. His book examined the development of a free society based on reason and justice.[4] The debating society at St. Andrews discussed Godwin's book heatedly. Some members of parliament wanted to ban it. Young as he was, Robert Gourlay dreamed of taking part in the formation of a society from which poverty and bad government would be banished. As his father listened to his son's talk and watched his character developing, he remarked, "Robert will hurt himself but do good to others."[5]

In 1794, when he was sixteen years old, Robert was initiated into national politics. He went to Edinburgh with his father to observe the trials of the members of the Society of the Friends of the People who were charged with treason.[6] He could not help thinking that Thomas Muir was correct when he declared his belief in the right of every man to have a vote and enough food to feed a family, but since the "Reform Martyrs" were accused of preparing arms to achieve their aims violently, he supported their deportation to Australia. The September Massacres in Paris, 1792, had shaken the minds of civilized men. Young Gourlay had a firm conviction that violence should not be used to achieve the perfect state that he believed was possible.

---

[4]In 1793, the year in which Britain declared war against the struggling French republic, Godwin published *Enquiry Concerning Political Justice and Its Influence on General Virtue and Happiness*. He hastened its publication as a protest against the banning of Thomas Paine's *The Rights of Man*, 1791.

[5]R. Gourlay, *General Introduction to Statistical Account of Upper Canada, compiled with a View to a Grand System of Emigration, in connexion with a Reform of the Poor Laws*, p. xcviii; *Statistical Account* II, p. xv.

[6]R. Gourlay, *Chronicles of Canada: being a Record of Robert Gourlay, Esq., now Robert Fleming Gourlay, "the Banished Briton"*, p. 6, from "Address to the Worthy Inhabitants of Niagara", April 2, 1818.

At the customary age of nineteen, Robert Gourlay finished reading for his M. A. degree. He tried no examinations at St. Andrews as proof of his scholastic attainment, for in the eighteenth century sons of gentlemen and sons of peers did not submit to such a plebeian practice. They merely passed on to further study. Many universities, including Oxford, still taught only classics and moral philosophy. St. Andrews taught natural philosophy (science) as well. Robert's interest in the latter had been whetted and he craved further knowledge. Already he was marked among his fellow students for his gay, free, honest manner and a bold and independent mind that might bring distinction to Fife, the shire of his birth. In 1797 Robert Gourlay entered Edinburgh University for more study in the course of which he would choose his vocation.

2

Edinburgh and England, 1800

During the summer months before it was time to attend classes in Edinburgh, Robert Gourlay visited estates of his father's friends in the Border shires to study husbandry under varying climate and soil conditions. Life was good and not too serious. He hunted on the Duke of Athol's estate. He indulged his young strength and love of nature by climbing Skiddaw, 3,300 feet above sea level, a giant compared to the hills of Fife. In the autumn, when he went to Edinburgh for university classes, he spent his spare time exploring the city. He deplored the crowding of wynds and closes among buildings built indiscriminately within the walls of the cluttered mediaeval town. He admired the draining of the Nor' Loch and the tidy planning of the new section at the foot of the citadel in which Princes Street was named after the numerous Hanover sons.

His main concern must be the choice of a vocation. Three professions—army, church or law — were considered suitable for young gentlemen of the time. The army did not attract him. He felt no call to the ministry like his friend Thomas Chalmers who had remained at St. Andrews where already the church was filled when he raised his voice to the Lord. He studied a bit of law under David Hume, but his father discouraged him from continuing because he himself had tired of the circumlocution of that profession. He was impressed by the profundity of Dr. Robison, a professor who had just published an eight-volume book on the dangers inherent in debating and corresponding societies.[1] He did not rush to join the classes of Dugald Stewart, the professor on the staff with a growing reputation for making students think independently. His primary interest was beginning to be the study of scientific agriculture, a subject so new to Edinburgh University that the first chair had just been established.

---

[1]The full title of Dr. Robison's book, 1797, was *Proofs of a Conspiracy against all the Religions and Governments in Europe, carried on in the Secret Meetings of Free-masons, Illuminati, and Reading Societies.* The Illuminati were supposedly preparing to undermine the government with a powder that would fill a bedchamber with pestilential vapours and a tea that produced abortions.

At the end of his first year at Edinburgh, he set out with a friend on another walking tour to observe agricultural methods, this time in the south-western Scottish shires. When they arrived in Wigtonshire, they were struck by their nearness to Ireland where the revolt of the United Irishment had just been suppressed at the battle of Antrim. In a youthful spirit of investigation, the two young men boarded a ship for Ireland.[2] They walked over the ground where the Irish had fought for freedom from absentee English landlords, freedom from restrictions on their religion, trade and industry, for universal manhood suffrage instead of the existing travesty of democracy. After two days the young men returned to Scotland to continue their holiday. Robert Gourlay expressed no disapproval of the severe repression of the United Irishmen. It was unthinkable that Ireland should be allowed to form a base for French attack against his country.

There was a certain justification for the aims of the malcontents, but it was wrong to resort to violence in order to establish a more liberal government, like the Irish; to protest food shortage, like the Meal Rioters; or to escape forced service in the militia, like the Militia Rioters.[3] It was true that the people had no votes with which to express their will, but young Gourlay considered them mad to revolt and destroy in the name of freedom.

The Tory government feared that Jacobinism was behind the rioting, and that George III's head might fall like that of the poor lockmaking king of France. In their zeal to vaunt their loyalty to George III, Scottish Tories would allow no Whig to hold a salaried government position. Edinburgh Whigs who wished to celebrate the birthday of Charles James Fox were met at the door of Writer's Court by the sheriff's men who held lanterns to their faces so they could record the names of the traitors. Some pulled their scarves over their faces and vanished before they could be recognized. Those brave enough to go in to the dinner discovered that the sheriff and Tory professor David Hume had their ears against the partition, eavesdropping for treasonous words.[4]

In 1797 an Act forbade all public meetings in Britain. In England public meetings were held at election time with the sheriff's permission.

---

[2]*Statistical Account* I, p. ccxvii; *Chronicles of Canada*, p. 6. (Omission of author and abbreviation of title will be used for second and succeeding mentions of works by Gourlay.)

[3]*A Specific Plan for Organizing the People and for Obtaining Reform Independent of Parliament to the People of Fife . . . of Britain!*, p. 58; H. Meikle, *Scotland and the French Revolution*, (Glasgow, 1912) p. 180 ff; Lord Henry Cockburn, *Memorials of His Time*, (Edinburgh, 1856) p. 82.

[4]Even the movement for the abolition of slavery was termed Jacobinical (Sir Keith Feiling, *A History of England* (London, 1963), p. 736) or seditious (Meikle, *Scotland and the French Revolution*, p. 78.

In Scotland they were non-existent. In spite of official zeal to suppress it, a dangerous rhyme made its way through the streets, "A wet winter, a cold spring, a bloody summer, and no king". Repression increased. There was no prospect of relaxation because the leaders of the government, William Pitt in England and Henry Dundas[5] in Scotland, said the danger from without was increasing. England's enemy was now Napoleon, for on the Eighteenth Brumaire Napoleon had supplanted the five-headed French Directory.

In the same year, at the end of his second year at Edinburgh University, Robert Gourlay chose his life's work. There was nothing earth-shaking about his decision. He simply decided to become a progressive farmer and leader in county affairs like his father. In time of peace a gentleman's son would have completed his education with a European tour to view the treasures of past ages and to enjoy some of the pleasures of the present that were best indulged in abroad. The continent was now closed to travel. Even Britannia's ships were excluded from the Mediterranean till Nelson won the Battle of the Nile with sailors who had just finished mutinying against their harsh living conditions. With foreign trravel out of his reach, Robert Gourlay returned to Fife to assume his responsibilities as a young laird.

The most compelling duty at the moment was to take up arms, for Fife had a long coastline facing France. In October 1799, he became a captain of the Volunteers but soon after there was a lull in the war. The French had not yet recovered from their naval defeat and Napoleon was preparing to negotiate peace in order to consolidate his position at home. The command of a troop of Volunteers was a flattering position, but it was not a full vocation for a young man with a questing mind. Fife was too small to satisfy him. Late in the year 1799, Robert Gourlay, accompanied by a Fife friend, set out to tour England and study scientific agricultural methods as he travelled.

As he passed through the best grain-growing land of the Midlands and East Anglia on his way to London, Robert Gourlay saw that much of it was still being cultivated with the wasteful open field and strip farming system of the Middle Ages. Tremendous tracts of land lay in an age-old state known as commons on which the villagers could pasture their stock free. The only way to make this land more productive was to fence it off for private ownership.

Enclosing had been proceeding slowly in England since the middle of the sixteenth century. Each enclosure had to be approved by an Act

---

[5] It was the pride of Henry Dundas to manipulate Scotland's voters so skilfully that in the 1802 election he delivered to London forty-three Tory supporters out of forty-five members. For this despotic control Boswell nicknamed him Lord Harry the Ninth of Scotland. *Ibid.*, pp.27, 33.

of parliament for this was communal land. If the process were accomplished by mutual consent of villagers and landowners, much good resulted from increased production. If the enclosing dispossessed the people ruthlessly, lasting hatred and hardship ensued, for independent cultivators became dependent on seasonal wages as farm labourers. Now that wartime fortunes could be made from the rocketing price of grain, landowners maintained that commons were nonsense in a country engaged in war and possessing a growing urban population with consumer needs. In twenty years prior to Robert Gourlay's journey to London 5,000 enclosure bills involving five million acres of land had been passed. Hundreds of English labourers eked a meagre existence on the poor rate with no hope of regaining their former independent status.[6] They sat in the ale houses whenever they had enough money for a pint: "If I am sober, shall I have land for a cow? If I am frugal, shall I have half an acre for potatoes? You offer no motives; you have nothing but a parish officer and a workhouse!"[7]

A change from old ways to new is often accompanied by misery. In 1799 the misery was calamitous, for it was the second year of poor crops in Britain. In the late wet spring of 1798 the seed rotted in ground that was as claggy as clay. The year 1799 was equally wet. In the first hard winter people sold their meagre possessions for food. This year they had nothing to sell. As he gazed out of the coach windows en route to London, Robert Gourlay saw the face of famine for the first time in his life. He could not force its horror out of his mind.[8] The young man who believed in perfectibility began to think as he had never done before.

Nevertheless, when he arrived in London the first thing he wanted to do was to view the sights, the greatest of which was the king. The Gourlays held the king in honour and gave him sympathy for his recent attack of mental illness. Robert stood outside a theatre to watch George drive up in his carriage. He went next to see Carlton House in the Strand to verify the tales he had heard of its extravagance. George Frederick Augustus, the heir to the throne, had built Carlton House with porphyry pillars, a great Gothic conservatory and a bathroom off his bedroom. Pearl-coloured, pearl-embroidered suits hung in his closets and Mrs. Fitzherbert was nearby. When parliament refused to pay his debts unless he produced a legitimate heir to the throne, he was good enough to get with child the graceless Caroline of Brunswick who was chosen for his wife.

---

[6]Twenty-eight per cent of the population was on the poor rate in 1802. (Ramsay Muir, *History of England*, II, p. 215.) Otmoor's change to private ownership was one of the most hated. "Oatmoor" was called after a mysterious lady of antiquity who had bequeathed to her people as much land as she could ride around while an oat sheaf was burning. Four thousand acres were fenced here. (J. L. and B. H. Hammond, *The Village Labourer*, 1760-1832, (Longman's, 1912) p. 88.

[7]Young, *Annals of Agriculture*, XXXVI, p. 508.

[8]*A Specific Plan*, p. 9.

Gourlay did not blame the king for the plight of the people. He blamed the royal advisers, the nobles who surrounded him in parliament, more intent on individual prestige than advising the king on the country's needs. He determined to visit the Board of Agriculture to ascertain what steps these advisers were taking for the relief of the famine.

The Board of Agriculture was a newly formed, semi-official body financed by a subsidy voted irregularly by parliament. Sir John Sinclair of Caithness had been chosen to be its head by virtue of his two important publications, a twenty-one volume *Statistical Account of Scotland,* 1793, a pioneer in the field of statistics, and the more recent *General View of Agriculture,* drawn up by counties. Because Fife had the highest rent value in Scotland (£3,872,600), Sir John made personal visits to many of the Fife improving landlords among whom the Gourlays were prominent. Sir John's answer to the critical food shortage was to urge the government to prepare a General Enclosure Bill that would speed up enclosures by making it unnecessary to apply to parliament for consent to proceed. Sir John had just enclosed his own enormous holdings in Caithness, in the process of which he "was under the necessity of . . . removing a number of ignorant and useless occupiers". Sir John believed that any evil attendant on enclosures was overbalanced by the good, for the statistics prepared by the Board indicated a steady increase in quarters of grain.

The secretary of the Board was beginning to disagree with the chairman's opinion of enclosures. The dedicated secretary was Arthur Young, author of *Travels in France,* 1792. Before the famine, Young had been a staunch supporter of enclosures. On his last trip to his estate in Suffolk he had been shocked to see the ragged bundles of humanity that huddled under the hedges, dispossessed by enclosures and refused parish relief by overseers of the poor who were trying vainly to keep the poor rate down. He had also observed that there were fewer poor in the parishes where the landlords had left the labourers a patch of land for a garden and a common for a cow. He was beginning to realize that the increase in enclosures was becoming a social mischief. He urged Sir John to include a clause in the new Act that would force every encloser to leave labourers enough land to pasture their cows, but Sir John had ignored him.[9]

The young man from Fife who had just come into his office shared Arthur Young's views about the effect of wholesale enclosures. "My father, and indeed my grandfather," Robert Gourlay said to Young, "had been in the habit of letting out small portions of land on a kind of perpetual lease, called in Scotland a 'feu', to labouring people whereon each man might build a dwelling-house, and enjoy the convenience of a garden.

---

[9]The agricultural labourer had lived well enough previously for the old people to remember good times. Now some said numbly, "All I know is that I had a cow and the government has taken it away from me." Hammond, *The Village Labourer,* 1760-1832.

I had marked the wonderful influence which the possession of such a small property had upon the characters of the people, giving them a superior degree of consideration among their neighbours, more steady habits, and more persevering industry. I had noticed with what serene delight a labourer, especially of the sedentary class, would occupy himself in his garden at hours not devoted to his trade, and I had calculated what an addition, as well to individual as to national wealth and happiness, such economical arrangements, generally adopted, might produce."[10] Young employed Gourlay to make a parish by parish survey of the condition of the farm labourers in the shires of Lincoln and Rutland.

Gourlay set out with zest. On December 7, 1799, he left London for Grantham.[11] With Arthur Young's letter of introduction he went from "pauper to Peer". In Belton parish where Lord Belton had left 100 acres as common pasture, the people were comfortable and cheerful. Some parishes refused to divulge their poor rate, but most cooperated. In the middle of February a large party dined with him at Brigg, Lincolnshire. In an excess of enthusiasm for his mission, Gourlay told them that the government intended to provide every poor man with land to keep a cow. A burst of approbation came from the company and all eyes flashed happily back at the speaker who promised a return to prosperity. This taste of leadership went to Gourlay's head. He had no doubt that the British parliament would take immediate notice of his investigation and there would be no more mobs in London shouting. "Bread! No King! Down with Pitt!"

Gourlay arranged his findings in a table that proved that in the parishes where the poor owned no cows because they lacked grazing land, the poor rate was as high as five shillings and eleven pence per pound. In the parishes that averaged one and a half cows per cottage, the poor rate was as low as seventeen and a half pence per pound. Arthur Young was pleased that the survey proved his premise, but Gourlay's small report impressed no one else. The Board of Agriculture even refused to allow Young to include it in its official report. No one was going to influence the landowners who dominated parliament to leave land in the inefficient hands of the poor. Gourlay was angry at the omission. To him the report did not represent cold statistics but human beings with names and needs. For the first time in his life he had encountered bureaucracy and government and his eager effort to assist in abolishing the widespread hunger of the people came to nothing.

Undaunted, he left London to travel through the south of England

---

10*Statistical Account* I, p. lxxxiii.

11*Ibid.*, pp. lxxxiii; xcvii; xci. *Annals of Agriculture*, 1802, XXXVII, pp. 514 ff. The table Young gave Gourlay may have come originally from Jeremy Bentham. Young, *Autobiography*, (London, 1898) Sept. 8, 1797, p. 308.

to assess conditions there. It was like entering a different country, particularly when he reached Wiltshire with its chalky, flinty soil. Enclosure had been completed here years before, for only large-scale farming could be successful on the thin soil of the downs and heavy bottom lands along the rivers. Here he met Thomas Davis of the Manor House, Fisherton de la Mere, who was passing around a Board of Agriculture questionnaire about agricultural conditions in Wiltshire. He readily exchanged ideas with the young farmer from Fife, for Scottish farmers had a fine reputation. News of the visit of this knowledgeable young Scot reached the ears of the Duke of Somerset who was buying up farms in the shires adjacent to his county seat at Maiden-Bradley in Somersetshire. He intended to invite some progressive Scottish farmers to manage them as soon as he was ready. For the moment, his recent marriage to socialite Lady Charlotte Hamilton and his mathematical studies into the elementary properties of the ellipse were absorbing his attention.

Nothing Gourlay had seen of the condition of the lower orders in the northern shires matched the concentrated misery he was viewing in the south. If some plan were not adopted to relieve the destitution, he feared that a new Wat Tyler would rise to demand action from rulers who refused to acknowledge that a crisis existed. Why should not he himself produce a plan to eliminate the starvation? All he needed was time for thought. He halted his travelling for a week and "set about the arrangement of the national economy", no small task for even a seasoned statesman. The first thing he must decide was the purpose of government. For whom did government exist—king, clergy, nobles or commons? His answer was the people. The majority of the people of England were agricultural workers. He divided the country mentally into portions and laid the foundations of agrarian law in what he called "The Village System": "for I thought that in societies of about a hundred families, all the conveniences of life could be procured with ease, and all the sweets of society sufficiently concentrated. I set my villages at such distances as I thought would not too far remove the labourer from his toil. . . . In front of each village I spread a green lawn, upon which the people's cows were to pasture. I gave to each house a portion of land for a garden; and I let the remainder of the soil to farmers, who were to pay their rents into a public fund, and contribute fodder for wintering the village stock. . . . I smiled upon the vigour and the virtues of my petty republics, and through the wisdom of primary and secondary representations, thought all domestic and foreign feuds could be silenced and averted. . . ."[12]

Many Utopians have gone no further than putting their ideas on paper. As Robert Gourlay neared home, he formulated a plan for his

---

[12] *A Specific Plan,* p. 11. Henry George is a later and bettter known proponent of the doctrine that all land rent should be paid to the state.

life during the next ten years. First he must prove himself as a successful farmer in Scotland. Then he would move "to England for a term of years, deliberately to study the cause of a difference so very great and manifest between the lower orders in the one and the other country".[13]

"Through life I have been enthusiastic in my pursuits," he wrote. Now he was ready to devote all his energy to his newly adopted mission. It did not matter to him that the task he had set for himself was full of menace, and that the poor had been in the world since the dawn of time.

---

[13]*Statistical Account* I, p. lxxxvii; cxxiii; *The Banished Briton*, #35, p. 481; #36, p. 491.

3

Laird in Fife

His new and intense conviction did not prevent Robert Gourlay from being glad to be home and part of the teeming, happy life of his father's establishment at Craigrothie. There was the long row of stone cottages with garden plots for the feuars, the kitchen garden walled against the wind, the carriage house, the stone wynd leading down the hill to the farm buildings, the mill on the stream, and the remains of a blacksmith shop built by an earlier Gourlay. There was the comfortable sandstone house that grandfather Robert had doubled in size by adding a reception room with a vine-carved recess, and a stone flying staircase that led on the one side to the bedrooms of the existing house and on the other to the drawing-room on the second floor of the new. Behind the house rippled the burn, and all around it were Scotch fir, fruit trees, and the larch that Oliver Gourlay had been planting at the time of his son's birth. Craigrothie House, "embowered in wood and in the season of bloom singularly beautiful", was a home to which the tall, dark-haired young laird was happy to return.[1]

While Robert had been absent in England, his father had purchased three more farms. He bought Pratis, east of Craigrothie on the Ceres-Leven road, for £2,800 in partnership with two others who soon sold out to him for £6,000. He paid £3,500 for Kilmaron on the Cupar-Perth road and would sell it soon for £20,000. He bought Glentarkie on the road to Abernethy on Tay. Pratis was the farm that Oliver turned over to his eldest son to manage. It was a good farm of 350 acres, all arable. The young laird loved the land. He instituted his new theories of scientific agriculture. He did everything he could for the contentment of his workers, despite the sniffs of older farmers who said he was daft.[2]

---

[1]Gourlay Papers, OA, 1852.
[2]*Statistical Account* I, p. clvii; *Statistical Account*, II, p. ix; *Chronicles of Canada*, p. 7; C. Rogers, *Memorials of the Scottish House of Gourlay;* Mrs. John Playfair-Hannay, Baltilly, Ceres, Fife.

Soon after he launched his course of improvements to Pratis, he received his next copy of Arthur Young's *Annals of Agriculture*. Young had published the results of Gourlay's recent survey in an article entitled, "Inquiry into the State of the Cottagers in the Counties of Lincoln and Rutland".[3] He was pleased when an English parson named Thomas Malthus referred to his article in the enlarged second edition (1803) of *Essay on the Principle of Population as it Affects the Future Improvement of Society, with Remarks on the Speculations of Mr. Godwin, M. Condorcet, and other Writers*. Malthus cited Gourlay's article as support for his own theory for abolishing poverty. On the contrary, Gourlay said as he discussed the book with other Fife gentlemen, he disagreed with Mr. Malthus. When Malthus considered the soaring poor rates, he said that poor relief should be abolished, for as long as the poor knew that they would get government hand-outs they would continue to have too many children. If the poor would curb their increase by practising "the law of virtuous restraint", there would soon be enough food for every one. Gourlay asserted that a parson who preached virtuous restraint to a man on his wedding day violated a law of nature. The dole and a lecture were wretched substitutes for the good life the lower orders could have if his system of cows for cottagers were adopted.[4]

By 1803 the Duke of Somerset was ready to rent his south of England farms to enterprising Scottish farmers. Twice he invited Mr. Gourlay of Pratis to occupy a farm in Wiltshire, but Gourlay was not yet willing to leave Scotland. Life in Fife was rewarding. He was a member of the exclusive Fife Hunt Club with its limited membership of one hundred. He wore with relish the required dress, a dark green coat embroidered with silver thread. His ready conversation made him a welcome guest at the best tables in Fife. He was celebrated for his pleasant disposition, the breadth of his knowledge and the energy with which he applied himself to his civil duties as son of a gentleman farmer and to his military duties, for he was of military age.[5]

During the year's respite from hostilities Napoleon had not relaxed. With incredible rapidity he began to make into fact the order, equality and progress that the revolution had failed to consolidate. For the first

---

3*Annals of Agriculture*, XXXVII, p. 514; XXXIX, pp. 240, 251; *Statistical Account* I, lxxxvii; *The Banished Briton*, p. 269.
4T. Malthus, *Essay . . . on Population*, Everyman edition, II, p. 245; *Letter to the Earl of Kellie*, p. 24; *A Specific Plan*, p. 20; *Statistical Account* I, pp. xcii ff. Gourlay praised Malthus' *Essay . . . on Population* because it was a contribution to the study of poverty, but he realized the shortcomings of Malthus' premise. Malthus' bogey of overpopulation would influence British currency, emigration, and poor law policy for over half a century. Some of the reformers, including William Benbow, already advocated birth control by contraception.
5*The Banished Briton*, #1.

time in a century of regal misuse of public funds, France was financially sound and feudal anachronisms had been swept away. Soon France would vote him consul for life with the right to choose his own successor who might be the son he was anxious to sire.

The British government feared "Général Vendémaire" for he was a threat to every hereditary crown. England had given asylum to the Bourbons and émigrés. French ladies walked in the streets of her southern towns with beribboned lambs on leashes, remnants of the costly simplicity with which Marie Antoinette had amused herself when the blue and gold palace at Versailles afforded no privacy. The confusion of wartime provided English nobles with a good excuse not to scrutinize government expenditures too closely. The country cried out about the corruption. Prime Minister Pitt himself resigned in 1801, charged with condoning government corruption, and his friend Addington took over. Within a few months Addington had bestowed a sinecure worth £3,000 yearly on his sixteen-year-old son.[6] "Pitt is to Addington As London is to Paddington", chanted the wiseacres in the streets who knew that Pitt's temporary retirement to the country would bring no change in government policy, no lessening of pensions and sinecures, no lowering of taxes, no more food for empty bellies. The government said that every energy must be devoted to the successful prosecution of the war.

Now Napoleon was massing ships at Boulogne. Fear of invasion closed in across the nearby island as chilling as the haar that blew in across the fields of Fife.[7] All over Scotland troops marched and turned and presented arms by daylight or nightly by the gusty light of torches. Walter Scott alternately composed cantos of *The Lay of the Last Minstrel* and drilled a troop to stab Frenchmen by setting up turnips on sticks. There were still those who did not find the war reason for holding their tongues. George Mealmaker, among others in Fife, seemed to have forgotten the fate of Thomas Muir and the Reform Martyrs of 1794. He invited men to his house to talk about parliamentary reform and maybe an end to the war. He was arrested, for United Scotsmen with subversive ideas could not be tolerated.[8]

Robert Gourlay was ready to prove his patriotism. He joined the Stratheden Troop of the Royal Fifeshire Cavalry, but he did not join as an officer. He joined as a private. This was odd, for he could afford the £400 for an officer's accoutrements. He was twenty-five years old. His tour of England had matured his keen powers of observation and developed his naturally critical nature. He had learned to scorn the conceit

[6]Pitt himself had refused the £3,000 sinecure of the Clerkship of the Pells. K. Feiling, *A History of England*, p. 725.
[7]Scots for a heavy mist that blows in from the sea over dry land.
[8]H. Meikle, *Scotland and the French Revolution*, p. 197.

of army officers and the corrupt practice of buying commissions. He was bold enough to say, "I preferred to evince my loyalty by entering as a private."[9]

His service in the Volunteers was short. The Whig opposition maintained that the Volunteers were useless in wartime as they could not be ordered abroad to fight. Suddenly, after forty years in opposition, the Whigs found themselves in office in December 1806. The first act of the brief Ministry of All Talents was to withdraw government support from the Volunteers. The Fife Volunteers met in Cupar to protest the insult, but Gourlay considered the action justified. Mounting the steps of the Tontine, he urged all the unmarried Volunteers to pledge to go to France if ordered. When there was no great chorus of Ayes at his suggestion, he resigned. He believed that if a war were to be fought, it should be conducted whole-heartedly, not played like a parlour game while it warped the life of the nation.[10]

He devoted himself to his civil duties as commissioner of supply. In the absence of elected county councils, county affairs were administered by heritors, justices of peace and three commissioners of supply.[11] The greatest need of Fife at the moment was new roads. Lately a few "outstrapalous" folk in Fife had dared to say that all the road money went to repair the roads that the lairds used most, particularly one that began and ended on the same estate. Gourlay knew that the scientific way to lay out roads was to have a contour map drawn up and to employ independent surveyors. (He had investigated the building of roads during his tour of the island.) One day he rode over to Wemyss Hall, a mile along the road from Craigrothie, to put his proposition before General Hope. General Hope said he would not support his proposals, and neither would the other lairds for jealousy was a characteristic of country gentlemen. The people of Fife continued to drive over roads laid out by "indolent, spiritless antediluvians", as Gourlay was beginning to regard the heritors.[12]

He had strong thoughts on Catholic emancipation as well. Roman Catholics had been excluded from office since the Test Act of 1673. Gourlay could not see why a Catholic should not be postmaster as well as a Protestant. The Tories opposed Catholic emancipation for fear of George III's displeasure. As a recent meeting of the Fife heritors, some of the county Whigs had placed the matter on the agenda of the next meeting,

---

[9] *A Specific Plan*, p. 45; *Appeal to the Common Sense, Mind and Manhood of the British Nation*, p. 34.

[10] *Ibid.*, p. 35; *The Banished Briton*, p. 9; *Niagara Spectator*, Aug. 6, 1818. Naturally, more was involved than the support of the Volunteers, the national issue being George's personal power. See " Reflections on the Dismissal of Lord Grenville's Ministry," CHR, June, 1964, pp. 84-104.

[11] *Addresses to the Electors of Fife*, Sept. 17, 1832, p. 8. Heritors were owners of a certain extent of land and their eldest sons.

[12] *A Specific Plan*, p. 15.

but the county Tories whipped in their members to a surprise meeting and voted it down. It was a brave Fifer indeed who dared to support Catholic emancipation, for the leaders of the nation and the leading men of Fife, the Earl of Kellie, General Hope, and Robert's own father, opposed it. Robert, however, boldly stated his belief in religious toleration.[13]

The whole spirit of administration was beginning to disgust him. He was beginning to "sicken at the surface spit and polish over the inner dirt of a society" that refused to come to grips with fundamental social and political problems such as poverty and religious discrimination. The life led by the hereditary leaders of the country seemed "hollow vanity". Now he stopped attending the Fife Hunt balls in his green and silver coat. He began to "ride about the county on a lean and ill-groomed horse: simpering aristocracy . . . grinned at the oddity; but it little knew that I constantly had in contempt the shallow respect which show and affectation can extort from the humbled peasant".[14] It was not as spectacular a method of drawing attention to wrongs as his ancestor Norman Gourlay had employed when he suffered martyrdom in 1534, but it was more than mere eccentricity. It was the gadfly method of stirring up action.

If "simpering aristocracy" was beginning to ridicule the behaviour of this young man, there were eyes in which he was finding the favour he desired. On March 8, 1807, when spring was brightening on the brae, he was accepted by the lady of his choice. She was Jean Henderson, daughter of a family friend, the late William Henderson, who rented Hatton Farm in the valley below Pratis. She was blonde, blue-eyed, vivacious and intelligent. Educated by a governess, she read and spoke French fluently, and her poetry was greatly sought after for inclusion in the scrapbooks of her feminine acquaintances. While Robert had been absent on his tour of Britain she had married John Stuart, the minister at Greenlaw. A year later she was widowed. On the day Jean accepted him, said Robert Gourlay, there was not a happier individual in the world. He would praise his wife openly: " I was the husband of as good a woman as ever lived."[15]

For the woman whom he took as his wife on August 10, 1807, Robert built a fine new addition to the old house at Pratis with its stair tower. It was now larger than Craigrothie House and comfortably Georgian in style. From the windows of the front bedrooms he and his bride could see Arthur's seat at Edinburgh across the Firth of Forth. From the side windows they viewed Largo Law looming over Hatton Farm in the valley below. Behind the house were the neat buildings of the renovated steading and farm cottages. In front stretched the lawn and garden that filled the house with the cool August scent of roses till the gloaming closed sweet and warm about them in the house that would shelter their love.

---

[13]*Letter to the Earl of Kellie*, p. 10; *Statistical Account* I, p. cccxxiv;
[14]*A Specific Plan*, p .15.
[15]*The Banished Briton*, #37, p. 500; Gourlay Papers, OA, March 9, 1837.

Upon the marriage of his eldest son, Oliver Gourlay contracted to give him an annuity of £300 and his wife half that amount if she should be left a widow. For any bairns grandfather Oliver provided a heritable bond of £4,000, interest not cumulative. This did not have to be distributed equally but could be portioned as the father saw fit. Thus was a Lear loop-hole left and the same power was given to the mother. Part of Jean Henderson's dowry was 866 acres of land in Dereham Township, Oxford County, Upper Canada, her share of 2,600 acres which the Honourable Robert Hamilton (an uncle who had gone to Canada in 1778 to seek his fortune with a Montreal fur trader named Ellice) had given to his widowed sister and her two children. In Britain a gift of so much land was the equivalent of a king dropping a ruby into the hands of a favorite, but at Upper Canada prices, sixpence a wilderness acre if a buyer could be found, the land would have brought about £20. This did not bother Robert Gourlay who thought eagerly of the challenge of cultivating rich virgin soil. He added an increasing interest in British North America to his already numerous enthusiasms, and he looked forward to the time when he and his wife would see their Canadian acres.[16]

A good life stretched ahead. "My father . . . stood preëminent for public spirit, and as a gentleman without stain. I and my family were provided for to utmost wish and beyond risk. . . . Out of doors, I was well with all ranks,—and in retirement blessed with the society of a most amiable wife, and her friend, the lady of Sir Henry Torrens.[17] At this moment, in spite of his dissatisfaction with county affairs, Robert Gourlay was happy, for all his mental, material and family needs were richly supplied.

---

[16]*Chronicles of Canada*, p. 7; Gourlay microfilm, OA.
[17]*The Banished Briton*, #4, p. 45. The former Sarah Patton, daughter of Admiral Robert Patton, nephew of Jean's father.

4

Adjourned While Speaking

Robert Gourlay's interest in Fife affairs encouraged his steady attendance at the monthly meetings of the heritors. On February 1, 1808, he set off with his father for the regular meeting in Cupar, the county town of Fife. This was an important occasion for Lord Crawford, the lord lieutenant of Fife, had died two days before at the early age of fifty. The important office of lord lieutenant would not be left vacant for long, for it was he who headed the Volunteers and knew which ones could be relied on to quell meal riots or militia riots, or to disperse illegal meetings like those of the Societies of the Friends of the People. Lord Crawford had been in precarious mental and physical health for some time. During the time of his incompetence his duties had devolved on the vice-lieutenant, the ninth Earl of Kellie. Lord Crawford's last breath had hardly sighed into eternity before Lord Kellie wrote to Robert Dundas (Henry Dundas' nephew and successor after his impeachment for corruption) soliciting his support for his elevation to the lord lieutenancy of Fife.[1]

There were few things Kellie wanted that he did not get. His boyhood had been spent in poverty, but his ambition and ability had enabled him to retrieve the family fortunes. An unexpected succession of deaths brought him the title he now enjoyed. In 1796 he had written to the Lord Advocate asking bluntly if his name was to be on the list of Scottish peers from which Henry Dundas "suggested" the sixteen peers whom he would like to see "elected" to represent Scotland in the House of Lords. In 1804 he swept grandly into the British House of Lords with the wine-red and ermine robes of a peer upon his shoulders. He was sixty-five years old, strong in frame and ruddy of face. He was near seniority in a shire in which every titled and landed gentleman knew his rank to the last triple decimal place.

As vice-lieutenant Kellie called the heritors' meeting to order, his

---

[1] Kellie to Melville, Cambo House, Jan. 31, 1808, Melville Correspondence, National Library, Edinburgh, Ms. 1048, p. 154.

thoughts were undoubtedly with the letter that was already bouncing on its way to London. Robert Gourlay also had more on his mind than appeared on the surface. The newspaper notice had stated that a discussion of the farmers' income tax was to be included on the agenda, and he had much to say on this controversial matter.

The first subject considered was the teinds (tithes) and higher salaries for ministers. The Earl of Rosslyn read seven resolutions, the gist of which was that ministers' salaries had been fixed in 1707 and they should remain fixed "for all time coming".

The meeting then resolved not to discuss the farmers' income tax. They would consider instead an address of loyalty to the king. Kellie's sense of timing was keen. An address of loyalty passed when he was in the chair ought to make a good impression on Dundas as he considered an appointment to the lord lieutenancy of Fife. The meeting decided to postpone the discussion of the income tax and adjourn so that the loyal address could be prepared.

Before the Earl of Kellie went home to Cambo House, he despatched another letter to Lord Melville in London written in his man-of-affairs script full of power and self-assertion. He wrote, and with truth, that he had been "vice-lieutenant for five years with the whole burden of the duty upon me at no small expense, that in point of rank I am the eldest peer residing in the country, that I have more political interest than all the peers together".[2]

A week later another missive was despatched to London, this time in the thin, crabbed writing of Lord Morton, Master of the Queen's Bedchamber. "I think it right to inform you that . . . I have stated to Lord Hawkesbury my expectation of being appointed Lord Lieutenant of Fife. . . . if contrary to my expectations, I shall be passed over, that circumstance will be a matter of triumph to a set of determined Jacobins who have taken a lead in the West of Fife."[3]

As the two contestants waited, Gourlay considered what to say about the income tax at the next meeting. The income tax had been imposed in 1799 by Prime Minister Pitt with great misgiving. In spite of taxes on luxuries such as carriages, men-servants, hair powder, horses, and all windows over seven in a house, there was still not enough to finance the war. In two years taxes had been doubled and tripled. Then the income tax was imposed at the rate of 10 per cent of gross income after £60 exemption.[4] In *The Wealth of Nations*, 1776, Adam Smith, the economist from Fife, had stated that he did not think it possible to impose a tax on income, but if such a tax were instituted, the government must allow ex-

---

2*Ibid.*, from Cupar.
3Morton to Melville, Park Street, Edinburgh, Feb. 15, *ibid.*, p. 162.
4A. Hope-Jones, *Income Tax in the Napoleonic Wars* (Cambridge, *1939*).

emptions for farm improvement. When the tax was announced, no allowance was made for exemption of the capital expenditures of an improving farmer, of which young Gourlay was a prime example.

Discontent with the first income tax regulations led to evasion. In 1804 Pitt inaugurated a National Revenue authority centred in London that sent travelling inspectors to supervise faint-hearted local collectors. If it was irritating to have a local man snooping into your affairs, it was outrageous to reveal your business to a cold and efficient stranger. The inspectors walked up even to commissioner Gourlay's door and asked, "How much rent do you pay? Do you have a manservant?" "Well, if you haven't, you can afford one and we'll tax you just the same."[5] In addition, the tax had to be paid in cash. At no time has the human race had enough money in its pockets. Now a farmer had to keep records and pay cash in a country that had not yet outgrown a barter economy and was resisting the change to paper currency. As he went around the county as commissioner of supply, Robert Gourlay listened to many complaints. The protests he wanted to voice at the coming heritors' meeting were more than the grand squeal of anguish with which any new tax has always been greeted. They were genuine grievances, but their consideration was being sought in war time when no discussion of government policy was countenanced outside the gentleman's club in London called parliament.

On February 15 the heritors again converged on Cupar. General Wemyss read the platitudinous address of loyalty. It was adopted unanimously for these were all loyal men. The matter of the income tax was then raised, but quick as a flash a counter-proposition was made to adjourn its discussion until April in order not to detract from the solemnity of the occasion.

Robert Gourlay had sat through more than one meeting that had failed to debate this matter. Rising to his feet he asked that they consider the income tax now, for the annual meeting in April was too crowded with routine matters to consider a question as involved as the income tax. Another man wanted to call a special meeting before April to consider the tax. Gourlay suggested that Thursday would be a better day than Tuesday because Thursday was market day in Cupar when all the tenant farmers were in town. The tenant farmers wanted to discuss income tax regulations too. By now all eyes were on him and many of the gentleman heritors were murmuring, "Tenant farmers really!" Gourlay should have taken warning from the disapproving looks, but his thoughts had been bottled up for so long that his words began to tumble out. The tenant farmers were no longer churls, he was saying. The tenant farmers have no way of meeting as we heritors have. Why should not we heritors take the lead in responsibility for the interests of those who cannot speak officially for

[5]*Letter to the Earl of Kellie*, p. 58.

themselves? *Why don't we invite the tenant farmers to the meeting to discuss the farmer's income tax?*[6]

Treason could not have produced a greater uproar than this last statement. Every one began to talk at once. The older heritors said that tenant farmers should be kept in their place instead of trying to usurp the prerogatives of those above them. Robert Gourlay's well-bred gentleman's voice grew louder in answer to the objections. He said he knew which proprietors were prejudiced against inviting the tenant farmers to the meeting and he was prepared to name them. In the midst of the "indecent clamour", the Earl of Kellie sprang from his chair and shouted that the meeting was adjourned. Gourlay roared back that he could prove what he had just said, but he no longer had the floor. The meeting had been adjourned while he was still speaking.

After some of the clash had died down, the earl came to discuss Gourlay's remarkable request with him. "You should consider yourself as one of us," he said, and a former member of parliament added, "The interests of landlord and tenant coincide and cannot differ," for he saw no reason for dissatisfaction with the *status quo*. According to them, Gourlay had turned against his class. If he had murdered his mother, he could not have roused more astonishment among his colleagues.

Robert's anger at being denied a hearing rose again when he read the report of the meeting in the *Edinburgh Advertiser*. It not only omitted mention of the argument but it also said that the meeting had concluded with a vote of thanks to the Earl of Kellie. Not only did the account lie, but it concealed the earl's arbitrary action. If the chairman could close an official meeting whenever he wanted, what hope was there for securing a free discussion of public affairs?

The question now confronting Gourlay was this: should he drop the matter and leave public affairs alone? He could not, for the issue at stake was the right to speak openly in what he considered should be a democratic body, not a feudal organization, and the necessity of correct reporting in the press. Three times he attended meetings of the heritors. Three times he was declared out of order when he tried to speak, once on the pretext that his request to be heard had been submitted in pencil, not ink. The man whose action he challenged was the man who desired to be Robert Dundas' right hand man in Scotland and lord lieutenant of its richest shire.

On July 27, 1808, Lord Kellie received his answer. Lord Hawkesbury, the home secretary, advised him that in view of recent incidents in Fife not he but Lord Morton was Lord Crawford's successor.[7] Robert Gourlay had been the partial cause of Kellie's loss of a coveted plum of office.

---

[6]*Ibid.*, p. 43.
[7]Hawkesbury to Kellie, London, July 27, 1808, Liverpool Letter Book, British Museum.

If Kellie suffered because of the hastily adjourned meeting, so did Gourlay. His stand on behalf of the tenant farmers was being discussed openly and people were beginning to take sides. Lords who always greeted him with customary Scottish camaraderie were snubbing him. Acquaintances crossed to the other side of the street in Cupar to avoid him. He tried to ignore it. He said that his crops "would grow in spite of it; and his friendships, if less numerous . . . are more select and warm," but with every day that passed he was made to feel an outcast because of the stand he had taken.[8] In ruder times he would have flung his glove at Kellie, but duelling was forbidden by law. He began to think constantly how he could clear his name as he talked to his friends or sat in the drawing-room at Pratis in the long evenings beside his wife. She occupied herself with her poetry writing, or reading the five volumes of Fanny Burney's *Camilla* that he had bought her the previous October, or Walter Scott's new poem, *Marmion,* or dreaming of the child quickening within her.[9] She also was feeling the effect of the criticism that was being directed against her husband. Some of her relatives had ceased to call on her for she shared her husband's conviction about the necessity for open discussion of public affairs.

Finally Gourlay decided to write to Kellie at Cambo House to ask for a written apology. Kellie replied briefly that as he was not aware that an insult had been made, he saw no reason to apologize. Gourlay then resolved to vindicate himself by publishing his version of the adjourned meeting in a pamphlet. In October 1808, nine months after the original incident, he published a sixty-three page pamphlet entitled *Letter to the Earl of Kellie, concerning the farmers' income tax, with a hint on the principle of representation.* It contained statements that revealed the temperament of a man who had sustained a personal affront, of a Scot who had more than his fair share of the thistle in his blood, and of a Fifer who was "a wee bit Fifish". It also made clear that Robert Gourlay's purpose in writing was related to the public good. He had plans for electoral reform but no prudent man in this time of censorship would throw out even a hint on electoral reform.[10]

---

[8]*Letter to the Earl of Kellie,* p. 10.

[9]Jean (Jane) Gourlay, born June 16, 1808.

[10]Sir John Leslie, liberal Whig professor at St. Andrews, wrote: "He never then appeared to think of general politics till his eyes were opened by the insolence of some of the Fife aristocracy, and his biting opposition to them was ridiculously magnified by such alarmists into a conspiracy against the state." *The Banished Briton,* #1, p. 12. See also Lord Henry Cockburn, *Memorials of His Time,* p. 78.

5

Letter to the Earl of Kellie

No one in Fife dared to publish Robert Gourlay's *Letter to the Earl of Kellie concerning the Farmers' Income Tax and a hint on the Principle of Representation.* No pamphlets on a political subject had been published in Scotland since the aborted flurry at the time of the publication of Paine's *The Rights of Man.* In any age in which man is mindful of his liberty, pamphlets have swarmed like flies. An age without pamphlets will mean that humankind is either at its lowest somatic level of nerveless acquiescence or that the choirs of heavenly perfection have taken over. No one in Scotland would publish his pamphlet, and it was not because perfection reigned.

No Voltaire was more determined than Gourlay to see his convictions in print. He refused to be silenced because Kellie controlled the Scottish press. He took his manuscript to London. Even in London, fear of government reprisal was great. It took him three weeks "to bolster the timidity of a wavering bookseller", for printers and booksellers were liable to imprisonment as well as authors. As soon as he could push it through the press, his pamphlet went on sale for two shillings.

First he gave a brief account of the adjourned meeting and Kellie's refusal to apologize. Then he revealed that the issue at stake was not personal pique only. "I lament from the bottom of my soul," he wrote, "the present frigidity of society, that torpor and indifference which seems to have taken possession of the minds of the worthy, that blind selfishness which can cheer the progress of distant rapine while it timidly sinks before the gnawings of domestic decay."[1]

Kellie was one of the selfish, wrote Gourlay. The class of which he was a member supported the continuance of an established church. This class clung to compulsory tithes "in the face of demonstration and famine, suffering tithes to weigh down the horn of plenty merely because in that

---

[1]*Letter to the Earl of Kellie*, p. 17.

26

posture it pours out provision for their younger sons".[2] This class had opposed the passing of the recent education bill. The Scots, wrote Gourlay, had been a barbarous people before the education bill at the beginning of the eighteenth century provided a school in every parish, a situation that made them better educated than the English. Now at the beginning of the nineteenth century the nobles in parliament had rejected an improved bill. "What did the rejection of the education bill prove? It proved that virtue and knowledge were making such rapid advances on the footsteps of vice that the rulers begin to dread the people".[3] This class was composed of landlords who considered tenants an inferior breed. Gourlay disagreed with them. Now that "every man can read, . . . every man may be a politician. . . . So changed are the times that . . . the present set of tenants are equally enlightened as the lairds".[4] These were the deserving men who were being refused entrance to the closed circle of the heritors' meetings.

Gourlay was tired of kowtowing to titles. Some of the nobility "whose blood has crept through scoundrels ever since the flood, can trace their descent from bastards of the royal family; and out of the heads of many which have not a single hair to bind them to the peerage, feudal folly cannot be eradicated".[5] How much respect would Lord Kellie command if he were still called plain Mr. Thomas Erskine, he boldly asked.

Kellie could not conceive how a man of Gourlay's position in society could jeopardize his position by adopting beliefs counter to accepted ones. He had told Gourlay after the adjournment of the heritors' meeting that "he should think of his own personal interest, and join with that class of men with whom government had decreed he should act for the public weal".[6] Gourlay saw that class "banded in a conspiracy against the rest of society . . . their only care that mind should not be cultivated, that no innovations should take place!"[7] Gourlay prided himself on his intense loyalty to the king, but he had only contempt for the nobles who surrounded him, whittling down the royal power by every means at their disposal instead of concentrating on good government to add to the nation's lustre. They perpetuated a parliament devoid of libertarian principles, for the fathers sat by right of birth in the House of Lords, and the sons by bought franchise in the House of Commons. Three hundred seats in the House of Commons were owned by members of the House of Lords. "There they have sucked in from infancy the too prevailing maxim that to cheat the King is no robbery".[8]

---

[2]*Ibid.*, p. 51. Tithes were sometimes collected with the aid of troops.
[3]*Ibid.*, p. 56.
[4]*Ibid.*, p. 55.
[5]*Ibid.*, p. 33.
[6]*Ibid.*, p. 4.
[7]*Ibid.*
[8]*Ibid.*, p. 42.

If Kellie would like to be carried with acclaim on the shoulders of the people (for he had often indicated his love of popularity), all that was needed to earn him plaudits would be to end the corruption in government circles. Gourlay treated the subject of corruption only in general terms. He could have filled volumes. In February, the same month in which he had protested the postponing of the discussion of the income tax, Sir Francis Burdett, a Whig who had deserted his class to lead the sansculottes of the borough of Westminster, had asked the Tories about £20,000 just granted to the Duke of York. The money had been seized from ships of countries with which England was not at war. He was told the money was the prerogative of the Crown. There was the scandal of the sale of army commissions by the Duke's mistress, Mrs. Clarke. There were the debts of the prime minister which his supporters excused for "he had no head for money" although he was the chancellor of the exchequer.[9] There were the enormous army expenditures that so far had resulted mainly in the disastrous Walcheren expedition or the defeat at Corunna. There were pensions for the rich but not for the poor.

The greatest example of all the government corruption had come in the person of Henry Dundas who lived with no great show of wealth, but who loved to bestow on his home Scottish folk coveted government and East India Company appointments with their opportunities for quick wealth. Henry Dundas, Lord Harry the Ninth of Scotland, first lord of the admiralty, had been in office so long that the public purse seemed like his own. He had advanced a friend's company £100,000 from admiralty funds though he knew it was near bankruptcy. Whitbread and Wilberforce proposed an investigation. Dundas was acquitted at the impeachment in 1806 but the stain left on his name was indelible. This was part of the corruption against which Gourlay inveighed.

Gourlay hated local bribery as well as national. In comparison with the kind that emanated from London, this was like guddling salmon, but one Fife man at least was beginning to object to being treated like a fooled fish. He resented the derisive condescension with which local parliamentarians treated the men whose votes they sought. General Scott, one of the two county members for Fife, "used occasionally to excuse his absence from Edinburgh by telling his friends that he had to cross the water to give his Fife lairds a feed!"[10] Kellie was good at trying the same methods. "Order a good dinner," wrote Gourlay. "Supply its deficiencies from your hothouse and your cellar where undoubtedly the wine is old and the hollands excellent. . . . These will make most of our homebred lairds perfectly sub-

---

[9]Feiling, *A History of England*, p. 725, considers that Pitt ended a great deal of the corruption. DNB says he died £50,000 in debt. For Dundas's impeachment see Eyck, *Pitt versus Fox*, (London, 1950) p. 356.
[10]*Letter to the Earl of Kellie*, p. 50.

servient."[11] Gourlay's standards of honesty were so fierce and high (some might call them unreasonable) that he hated even the custom of free political dinners, designed in this case to secure the five votes commanded by his father.

Governing a country, he considered, should be regarded as seriously and cleanly as practising religion. There was little room in Gourlay's political thinking for a consideration of that complicated grey world between black and white where decisions have sometimes to be made to a politician's sorrow. What would he do to stamp out government corruption? He would convene all the honest men in Fife who could read and write. They would swear to preserve order, not act like a mob, and would vote on public questions by a well-secured ballot. The first resolution that they would pass would be "that bribery and corruption are worthy of death".[12] They would send this to parliament immediately so that it could pass a law to that effect.

These were the "hints" that Gourlay published about the political system perpetuated by the unrepresentative parliament of his day. As for the income tax, he said that Kellie had never intended to discuss it. He had only advertised it on the agenda to entice a large attendance to indicate that he was well supported in his candidacy for the lord lieutenancy. The arbitrary adjournment of the meeting insulted not only Gourlay but also all the worthy tenant farmers who had no way of making themselves heard.

The pamphlet, *Letter to the Earl of Kellie,* was no marvel of organization or mellifluence. It was a maiden effort written by a man who was tired of empty oratory and speeches filled with truisms. He wanted to shock people into listening to him, and though his tone was easy and respectful throughout, every sentence was barbed. He referred to the French farmer with only a sentence, but he knew that one reason for the war was that the progressive French farmer produced cheaper grain than the English. He knew that another reason was to maintain the merchant members of parliament like Lord Crawford in the West Indian carrying trade.[13] He knew the men who sold their votes to procure a cadetship in India for their sons. He knew those, like Lord Morton, who opposed the abolition of the slave trade. He knew both local and national affairs, and his pamphlet touched every current controversy with a tongue like fire.

What, then, did Robert Gourlay propose as a cure for the diseases of the body politic beyond a meeting of honest men to forward a resolution to

---

[11]*Ibid.,* p. 32.

[12]*Ibid.,* pp. 34, 50. Not till 1883 would Britain pass the Corrupt and Illegal Practices Act.

[13]The carrying trade was so important in world economy that the slump caused by Napoleon's embargo brought the depression of 1808. This resulted in the unrest that occasioned Gourlay's outburst in Scotland and a flare-up that will be noted later in Upper Canada.

parliament to make death the penalty for corruption? The only remedy lay in the reform of the individual. The standards of private morality must be applied to public actions, and strict Scotch Presbyterian standards at that. If "revolution may be effected without an offering of blood . . . My lord, there is but one way of accomplishing it with safety; and that must be by our efforts as individuals. We must subdue our passions . . . we must sacrifice our private feeling to the necessities of the public".[14] It was the solution of an idealist. It would be easier to wrestle with the winds of winter than to expect the men who were engaged in the greatest game of all—pulling the strings by which a country moved—not to make under-the-table deals, not to flex their restless mental muscles in planning financial and martial campaigns on a global scale. Most hopeless of all, it was the solution of an idealist who would scorn the crooked reed of party politics even though it was the main staff available. Because in his opinion both parties required reforming, Robert Gourlay would lean on neither.

Though Gourlay disclaimed adherence to any party, he was close to the views of an almost forgotten party called the Country Party which opposed court corruption and tried to ally King and Commons against baron and userer.[15] They both had a romantic disregard for the fact that the king was George III, pestered by greedy sons, quarrelling daughters and seekers for pelf, troubled since 1788 with periods of insanity, and very Tory in the upper storey when he was sane. Cabinet government had not yet evolved into maturity. When the cries from the country, "Change your ministers", were loud enough for him to hear, the king would change his ministers and then call an election to approve what he had done. Even if the king had been different, the Country Party had no chance of electing members for all the seats were in the buying hands of the Whigs and Tories. By the time *Letter to the Earl of Kellie* was published, Pitt the Younger—the man who maintained that the king should choose his own ministers when others like Fox were already claiming that they should come from the party in power in the House of Commons—Pitt had been dead for two years, but his successors and the king saw no reason for change. In the middle of such a situation Robert Gourlay, writing from a country still regarded by England as barbarous, raised his voice with a solution for ills that the government did not even acknowledge. Certainly his pamphlet contained as well a modicum of a sense of personal insult. "Your act, my lord, confirmed by the heritors of Fife, has deprived me of my birthright—of that dignified right which made me feel of some consequence in my native country".[16] It contained a hint of the "sour grapes" of a man who may have spoken from personal envy of aristocrats who had been honoured with titles for services

14*Letter to the Earl of Kellie*, p. 29.
15A. S. Foord, *His Majesty's Opposition*, 1714-1830, (Oxford, 1964).
16*Letter to the Earl of Kellie*, p. 41.

no greater than those of his own ancestors. It would not have been written if Kellie had not considered himself too grand to apologize. "It is a melancholy reflection," wrote Gourlay, "that they who hold power and privilege . . . should be so tenacious of it, so jealous, as to fly off from the slightest suggestion of proffering its assistance . . . in a liberal way to others".[17]

Three months after Gourlay set out with his manuscript for London, his printed pamphlet reached Fife. Its reception was mixed. From some, wrote Gourlay, he said he had "the best satisfaction which cheerful faces can give, that my *Letter to the Earl of Kellie* was well received". From others, such as the Earl of Leven, "I have been told that trifles can be construed by the law into libel".[18] Robert Gourlay, who as a youth had pondered the abstract problem of perfectibility that William Godwin had discussed in the safe sphere of theory and examples drawn from antiquity, was now the man who was fighting reality in the form of living government figures. Antiquity was voiceless, but the present was full of tongues. The leaders of Fife instructed the sheriff to seize Robert Gourlay's pamphlet and commence an action against him that could lead to jail or deportation.

The suit was never executed for he was preparing to move his wife, baby Jean, his furniture, farm implements and ploughmen to the south of England to one of the Duke of Somerset's farms. He had served his apprenticeship as a farmer in Fife, and now the time had come for him to enter upon phase two of his life's work, the study of the poverty of the English farm labourer. While he had been in London superintending the printing of the pamphlet that had raised such a storm, he had made a tentative agreement to occupy seven hundred and fifty-acre Deptford Farm, Parish of Wily, Wiltshire. He had one more trip to make south for final arrangements about the lease before his departure.

He was also busy writing a second pamphlet far more incendiary than the first. The Kellie pamphlet had merely hinted at the reform of parliamentary representation. The manuscript Robert Gourlay carried to London in the spring of 1809 was explicit enough to make the Fife nobility congratulate themselves for unloading this man onto England.

[17]*Ibid.*, p. 51.
[18]*A Specific Plan*, p. 5; *The Banished Briton*, #12, pp. 105, 106.

6

People of Fife! . . . of Britain!

As he journeyed to London, there were no regrets in Robert Gourlay's mind at the prospect of leaving his privileged position in Fife. "Blessed adversity! I could almost worship thee! Grateful am I for the disdain of the world; for through contempt of it I feel myself free. Thankful am I for moderate prosperity for its closer embrace might have clogged me with acres, or cursed me with care," he wrote in his second pamphlet, *A specific Plan for Organizing the People, and for Obtaining Reform Independent of Parliament—to the People of Fife . . . of Britain!*[1] This was the thinking that changed his attitude: "The structure of my thoughts has grown up under the most peculiar circumstances; and from the wreck of the most liberal education all that I can feel as saved to me of value are *certain principles*. At school, and at college, I heard nothing but the purest independence inculcated. It was said to be the noblest privilege of man;—it was understood to be the peculiar prerogative of Britons: but I am now in the world, and all experience tells me that hitherto all its practice had been only among boys."[2]

Though his boyhood idealism was shattered, Gourlay still drew strength from the protesters of history. He extolled Junius, the anonymous writer who had dared to attack the corruption of the "King's Friends" forty years before. He praised the founders of the United States of America: "People of Fife! see the rotten eggs of Britain thrown ashore on the continent of America come alive! . . . Think of their shop-boy Franklin;—this day wandering contented with a dry roll; next the ambassador of America to France."[3] He dared also to admire Thomas Paine, though since 1793 the penalty for printing and selling his books was imprisonment. "From the earliest period of my recollection, I had but one feeling as to the natural

---

[1]*A Specific Plan*, p. 135.
[2]*Ibid.*, p. 2.
[3]*Ibid.*, p. 25.

rights of man: and I needed no lights from Thomas Paine, . . . to convince me that governments had universally been the sole promoters of war, and rulers the chief tormentors of the people;—that the British constitution (I confine myself to what I dare, the representation of the people) so far from being venerable, was a mere concentration of power, a nonentity imposing on the fluttering spirit of liberty."[4] Though Paine was a deist republican and Gourlay a God-fearing monarchist, he could see no reason why many of Paine's logical ideas of reform should not be adopted.

Most of all, Gourlay admired William Cobbett, the self-educated gardener's son who had begun his political life by opposing the ideas of the French and American revolutions.[5] In 1802 he hated France so much that he had refused to light candles of rejoicing in the windows of his Pall Mall bookshop when all around him were celebrating the Peace of Amiens. He was made editor of a government-subsidized newspaper and was allowed to share some of the privileges reserved for government supporters. When a war loan was being floated, one of his government friends wrote out a pledge for a sum from Cobbett. Before he had a chance to pay, his finance-wise friend said that the demand for government bonds had raised the price, and handed Cobbett £200 profit.[6]

This led the scrupulous Cobbett to study Thomas Paine's *The Decline and Fall of the English System of Finance.* He began to scrutinize the activities of the stock jobbers who had mushroomed since the beginning of the war and the Acts on loans and banking that gave politicians the advantage over the general public through manipulation of public funds. He began to question Pitt's celebrated sinking fund that his friends hailed as a stroke of genius, but which Cobbett thought was like sending out a cow to catch a fox. He began to connect the rise in the number of paupers and the poor rate with rack-renting[7] landlords who knew that they could profit by the fluctuation in the value of money produced by what was elegantly known as high finance. He was no longer simple enough to think that the price of grain depended solely on the size of the crop, and that all government expenses were legitimate. At the age of forty, Cobbett's study of government finance changed him so completely that he turned against both parties. His paper, the *Political Register*, became the voice of the people, of a party which as yet had no name and no representation in parliament.

This was the man whose style of writing Gourlay was adopting. Cob-

---

[4]*Ibid.*, p. 8.
[5]*Ibid.*, pp. 6, 96.
[6]G. D. H. Cole, *Life of William Cobbett*, (Collins, 1924) p. 85. When Arthur Young was handed a similar £200 profit, he liquidated a few debts saying, "I was thankful to God for this and meditated much on it. . . . If God pleases to give me money he has a thousand ways of doing it." *Autobiography*, p. 379.
[7]Yearly leasing instead of long term.

bett employed any language he thought would secure him a hearing. "I have heard him reprobated for vulgarity and abuse," wrote the well-bred young man from Fife.[8] Nothing could be truer. When Pitt died in 1806, his government friends immortalized him with interment and a statue in Westminster Abbey. There stands his statue still amidst a marble forest of the great, but few who view it now hear the angry voice of Cobbett who opposed Pitt's policy, his suspension of habeas corpus, his Treason and Sedition Acts, his funding system and the people's poverty. Even after death Cobbett did not spare him. Nobody regretted his passing, he wrote, except "the bloodsuckers and muckworms" of government circles.

Those "bloodsuckers" were the men who were threatening Gourlay in Fife with a libel action, and they were the men who were preparing to arrest Cobbett himself. They were the men whose unchanging ways were causing pens like Cobbett's and Gourlay's to drip vitriol. Since Sir James MacIntosh had written his reasonable *Vindiciae Gallicae*, seventeen years of war had passed, seventeen years of quadrupling taxes, of enclosures, of factory town misery, of dying soldiers, of deportation, of repression. Only a short time before, Hawkesbury (Lord Liverpool) had written to Sir Arthur Wellesley (the Duke of Wellington), "I trust that you will succeed in keeping the country quiet without having recourse to the Insurrection Act. I will send you more troops if necessary."[9] Because sane arguments brought no redress, Gourlay wrote: "Nowadays it is a duty to be vulgar and abuse: chastisement must be fitted to its object;—the filthy toad can only be kicked from the foot-path."[10]

His country needed a leader to save it from the morass into which it had sunk. The leader might come from his beloved home shire of Fife. What were its people like? "We are men, and we profess to be Christians. Do we exercise the *rights of men* and keep in view the *duties of our religion?*"[11] Was there a leader among them? Gourlay's previous pamphlet had given his opinion of the Earl of Kellie. Now he expressed his opinion of some others. Those among the Fife nobility who had secretly rejoiced at Kellie's discomfiture after the publication of *Letter to the Earl of Kellie* had laughed too soon. No such inclusive castigation as Gourlay's second pamphlet had appeared in Scotland since Sir John Scot's *Staggering State of Scotch Statesmen* in 1754.

Would a possible leader for Britain from Fife, asked Gourlay, be the Earl of Morton who, on the last Queen's birthday had "to shrink under the petticoats of the queen, and call her to seize the purple mantle from the

[8]*A Specific Plan*, p. 7.
[9]Hawkesbury to Wellesley, London, October 22, 1807, Liverpool Correspondence, British Museum.
[10]*A Specific Plan*, p. 7.
[11]*Ibid.*, p. 1.

marble table, in order to strengthen his barricade from insult?"[12] Would it be the Earl of Leven who at the last Hunt Ball had asked Gourlay to lie by bidding up the price of his Devonshire bull at the next auction, saying that he would buy the bull back if Gourlay were left with it? Would it be an important Fife ecclesiastic who had made his fortune by gambling? Would it be the Earl of Crawford, "a thing lost to reason, and lost to modesty?"

Or would the Fife leader of Britain be General Hope who had just published his version of the glorious victory at Corunna, ignoring the fact that the walls of old England had been forced to evacuate in two days an expedition it had taken the government months to prepare and weeks to get ashore? Gourlay had gone down to Portsmouth himself to check on the truth of the flattering reports. True enough, General Hope was guest at a feast in his honour, but Gourlay heard from the wounded lying on the stony beach the true story that Lord Castlereagh, Secretary for War and the Colonies, was afraid to give to the public. More than that, Gourlay knew the falsity of war itself: "When labour is thrown away upon war, which never obtains its object, it is gone forever, leaving no ransom but misery".[13]

Gourlay maintained that his exposure of the leaders of Fife was not made "in the ebullition of the moment. . . . The people of Britain will never be aroused by abstract lectures: they must have reference to living characters;—they must see with their own eyes . . . the demons that are disturbing their peace." The government must be awakened to the need for reform before Britain, like France, required a revolution to cleanse herself. Change should not be considered dangerous but salutary, and the first changes must come from the aristocracy whom Burke called the Corinthian pillars of a polished society, but whom Gourlay called self-seekers around the throne. "O the times! for they have betrayed even good men!" Gourlay lamented as he considered the difficulty of effecting a transformation. "O places and pensions, ye are irresistible! O aristocracy! thou hast degraded our king; thou art not only the gloomy ivy which feeds on the mouldering walls of palaces and lordly towers; the crawling moss which adheres to every stunted plant stuck in an acre of feudal ground art thou!"[14]

---

[12]*Ibid.*, pp. 27-36. During the periods of the king's mental illness, some nobles supported the queen for regent and some the heir who disliked his father so much that he had fleurs de lis woven into one carpet at Carleton House when George III renounced the English claim to the French throne in 1801.

[13]*Ibid.*, pp. 42 ff.; p. 85. For years school children were taught the glories of war:
"Slowly and sadly we laid him down,
    On the field of his fame, red and gory,
    We carved not a line and we raised not a stone
    But we left him alone in his glory."
                        "The Burial of Sir John Moore", by C. Wolfe

[14]*Ibid.*, pp. 7, 28.

The disgraceful national situation that he described so rhetorically could be remedied by a reform of the faulty system of parliamentary representation. Gourlay knew firsthand the ways of the parchment barons of his own Scotland.[15] They were a bagatelle compared with England's rotten and pocket boroughs. Trusting no man's eyes but his own, he had gone to look at the notorious rotten borough of Old Sarum, a depopulated ruin of a cathedral and fort outside Salisbury. He had gone to see Gatton, one of the fifty-nine "scot and lot" boroughs in England, survival of the day of James I when every householder who watched and warded for the king could vote. Gatton had six houses, one voter, and two representatives named by the owner of the land. He had seen part of a disgraceful English election at Stamford one midnight when the coach stopped there on his way back to Scotland. Every one in the inn was shouting, "Oddie and liberty restored", and drinking freely of Oddie's liquor. In these days before secret balloting, each candidate tried to keep the other man's supporters too drunk to reach the poll and his own just drunk enough to be happy. This equilibrium was difficult to maintain, for elections sometimes lasted three tumultuous weeks of bribery, black eyes, cracked heads and rioting. The whole system was scandalous, according to Gourlay, and should be changed.[16]

Some men like Colonel Wardle asserted that parliament should reform itself. Gourlay said that Colonel Wardle was wrong to expect the present parliament to reform itself. "It makes me think of a scavenger falling down before his worn-out broom praying to its wasted stumps to put forth new shoots, and repair its original vigour and utility."[17] Who in his right mind could expect a parliament composed of two corrupt parties to reform itself? The proper way would be to elect a new parliament by means of a wide franchise and ask the king to declare it the new, clean government of Britain. Gourlay was finished with hinting; he had a full-fledged plan to present.

First were the qualifications for a voter. Property as a requirement was outmoded. Every nineteen-year-old male who could read and write should possess a vote. Second was the size of the polling district. Three hundred voters was the ideal size. Third was the frequency of elections. They should be annual, so that voters could check their representatives closely.

Fourth was the method of voting. Voting should be conducted on one

---

15"In the county of Fife, the property which qualifies for a *real vote,* upon the average, will cost not less than 20,000 l. A *fictitious vote* may be bought for 400 l." *Letter to the Earl of Kellie,* p. 52.
16*A Specific Plan,* p. 173; Gatton's voting peculiarities from Bentham, **Plan for Parliamentary Reform,** 1817.
17*Ibid.,* p. 173.

day in order to prevent brawls. Because indirect elections prevented mob-
bing, there should be primary and secondary elections for representatives.
At six o'clock on May Day morning voters would assemble, prove before a
minister that they could read and write, and then file silently into a hall
where they would record their votes secretly by means of a voting machine.
All this could be completed decorously by eleven o'clock in the morning
when the fortunate representatives of the primary polls would be declared
elected. These men would then proceed to a central meeting-place, say at
New Inn for the east of Fife. There they would elect the final representatives who would go to London to cleanse the Augean stables of the present
government. Their expenses would be paid by the shires. Near London
they would convene with similar representatives from the rest of Britain
at the house of someone favourable to reform, such as Samuel Whitbread,
(Grey's brother-in-law). They would petition the king for an audience, and
he would proclaim these incorruptible men to be the parliament that
would bring good government to the land.[18]

Gourlay anticipated some of the obstacles that stood in the way of his
plan, such as a hall for the voting. The only places in Fife large enough to
accommodate three hundred voters were the parish churches. Gourlay suggested that the pews in Ceres church, a mile from Craigrothie, could be
rearranged so that the voters would face a mechanical ballot box placed in
the centre of the church. This ballot box would ensure the secret balloting
indispensable to true democracy. He had not yet worked out all the
details of the ballot box with three hundred compartments, but he planned
to ask an ingenious carpenter to help him. To ensure secrecy he would
place his box behind a "veil".[19] It was a good plan—on paper.

Scottish people were accustomed to listen to four-hour Sabbath sermons
and to read the Bible for the remainder of the day, but Gourlay asked all
non-voters to remain inside on election day reading the laws of their
country. Not since the day of Lady Godiva had any populace remained
indoors for a similar length of time, and even the preternatural hush of a
Scottish Sabbath afternoon, when the very cattle seemed to crop more
silently than usual, was breached by the dogs who waited outside the

---

18*Ibid.*, pp. 20, 78, 312.

19*Ibid.*, 111-114. The problem of secret balloting had not been solved by Athenian
ostrakons and black balls. Contemporary efforts to solve it are recorded by Edinburgh
Reviewer Sidney Smith, *Selected Writings*, ed. W. H. Auden, (Farrar, Straus and
Cudahy, 1956) p. 347. Historian George Grote exhibited a dagger ballot box around the
country. The voter stabbed his candidate's card with a dagger. Another, called the
mouse trap, grabbed the finger of the voter till a trap clerk pulled a liberator to release
it. The eminent mathematician, Charles Babbage, experimented with a mechanical
ballot box in conjunction with his research into a calculating machine. Secret balloting
would not be adopted by Britain till 1872.

churches for their masters. It was not likely that the populace would add a day of civic solemnity to the quota set by their religion. In addition, the people of Ceres parish were aghast at the suggestion that the pews of their church be turned around so that a monstrous secular ballot-box could be placed near the sacred psalm-singing berth.[20] They had just paid for renovating the church (1805).

There were other objections too. The English were angry because Scotland's higher rate of literacy would give Scotland a greater proportion of voters. The country gentleman feared he would lose his importance if ridings were made equal and every hind had the same vote as he. Indirect election smacked of the constitution of the hated American republic. The biggest obstacle of all was the king whom Gourlay revered. Farmer George understood the workings of his prized Spanish rams better than a ballot box or the forces rising in his country. He was convinced that democracy and revolution were synonymous, and either would lose him Hanover. Besides, not all the people whom Gourlay hoped to benefit through electoral reform disapproved the high jinks that accompanied unreformed election times. A break in routine was always welcome, and in London recently the people of Porridge Lane had fought to grab the bread and cheese thrown to them by the Duke of Northumberland's liveried servants. Election customs were acceptable to approximately the same number as they offended, but to the rigidly upright Robert Gourlay they represented bribery that must be eliminated.

When he arrived in London, in May 1809, with his manuscript entitled *Plan for Organizing the People and for obtaining Reform Independent of Parliament—to the People of Fife . . . of Britain!,* Gourlay again had to search for a courageous printer. While he waited, he tried to contact the Duke of Somerset in order to sign a lease for Deptford Farm. He called at his house in Park Lane but the duke had gone to his country seat at Maiden Bradley. He journeyed to Somerset but the duke had returned to London. He looked over the farm in Wiltshire and returned to London to discover that the duke was elsewhere.

Now he learned that a dinner was to be held at the Crown and Anchor tavern to celebrate the second anniversary of the election in the borough of Westminster of the first two reform members in the British parliament. The dinner drew him like a magnet for in all his thirty-one years he had never attended a public meeting because of the rigid enforcement in Scotland of the law against them. An open meeting such as this dinner at the Crown and Anchor was frowned on in England but not forbidden, for the men whom it was to honour were two aristocrats, Sir Francis Burdett and Lord Cochrane. Their election campaign had been managed by self-

---

20M. Conolly, *Biographical Dictionary of Eminent Men of Fife*, (Cupar-Fife, 1866).

educated Francis Place from the book-filled room behind his tailor shop at Charing Cross.[21]

The Crown and Anchor possessed a dining-room large enough for the celebration of the victory that was the thin edge of the wedge that would result in the Reform Bill of 1832. Gourlay was a little intimidated at the crush of strangers, but he had taken a written speech that he hoped to give in order to identify himself with the friends of reform. He was disappointed to discover that the only speakers would be those at the head table. Sir Francis' speech was more moderate than he had expected, but he joined the loud "Hear! Hear!" when Burdett declared, "Indeed we belong to no party but that of the people." The speaker with whom he felt the most affinity was Major John Cartwright, one of the mainstays of the Corresponding Societies so feared by Pitt and Burke. Clamp-mouthed and square of chin, this man was no theorist. He had simply closed his jaws on the idea of parliamentary reform and never let it go. Gourlay resolved to meet Major Cartwright as soon as he could to compare their aims for parliamentary reform.[22]

He compensated himself for not being able to speak by adding a postscript to his as yet unpublished pamphlet. He described the great political meeting to the people of Fife who had never attended such a gathering. He criticized it too. Why were the London leaders so slow in organizing Britain for reform? Why had not Burdett, Cochrane, Waithman, Cartwright and Place declared that they would sit at the Crown and Anchor the very next day at ten a.m.? In one day they could divide London into four hundred polls of three hundred voters each. By the king's birthday on June 4 the final candidates could be ready to approach the king and be declared the incorruptible parliament of Britain.

Gourlay's motives were pure and his proposal was not necessarily incendiary. England had changed its government without violence during the Bloodless Revolution of 1688 when a convention summoned a new king who called a legal parliament. Disgust with parliament's refusal to reform the system would soon cause the city of Birmingham to elect an unofficial representative. Cobbett too would make a proposal for a "parallel parliament" to call parliament to mind its duty to the country. But Cobbett had just been arrested and jailed without trial for his article protesting the use of German mercenaries to flog the British soldiers who had refused to shoot at the people rioting for bread. Now Gourlay received

---

[21]*A Specific Plan*, pp. 156 ff. The borough of Westminster had an electorate too large to bribe. G. Wallas, *The Life of Francis Place*, (Knopf, 1918) Chapter 2. The Crown and Anchor, on Arundel Street just off the Strand, was the headquarters for the reformers, Brooks' for the Whigs, White's for the Tories, and Boodle's for neutral country gentlemen.

[22]*A Specific Plan*, p. 164.

a letter from his wife in Scotland urging him to be careful, for the sheriff was again being pressed to arrest him. He finished describing his scheme for a reformed parliament, then added defiantly: "These are my opinions—my principles: and by these I care not though I am tried, in Scotland or in London".[23]

Not every one who was convinced of the necessity for the reform of parliament possessed the same courage as Gourlay when he published this pamphlet. In this year of 1809 Jeremy Bentham wrote a pamphlet called *Plan for Parliamentary Reform in the form of a catechism* . . . . Bentham issued from his retreat off Queen's Square to look for a publisher for his manuscript. He did not persist till he found one. He took it back to The Hermitage where it lay for eight years. Bentham, his philosophical friends such as John Stuart Mill and his political friends such as Henry Brougham, called people like Gourlay and Cobbett "obscure desperadoes with the sacred name of reform on their lips". Bentham declared that the road to national ruin began in the year 1809, but he did not insist on having his pamphlet published that year as a step on Britain's long road to reform.[24] When Gourlay decided that the time had come for him to act, he found a printer.

Gourlay's second book was no more a marvel of controlled organization than his first one. He pelted his readers with his ideas on reform of corn laws as well as parliament. He teased the poetic senses with rich phrases like "the dangling, antick hieroglyphics". He tossed out a whole social philosophy in a sentence. "Every society is bound, *in honour,* to take care of certain unfortunates; it is the business of society to reduce their number as much as possible by fair means: whatever happens in the world, the abstract virtue of this law must remain the same".[25] He was the eighteenth century gentleman displaying his knowledge as well as a nineteenth century reformer outlining his plans for betterment. He was still the stern Presbyterian from Fife setting forth impossible standards of truth and moral perfection. "Write no man's epitaph for honesty till death has finished action", he concluded, thinking of Henry Dundas who had broken Scotland's heart when he was accused of misappropriating his country's funds.[26]

Gourlay returned with his second pamphlet to Fife. He was not clapped into jail because the sheriff knew of his plans to move to Deptford Farm,

---

23*Ibid.,* p. 155.
24J. Bentham, *Plan for Parliamentary Reform,* p. 1. Cobbett called Bentham "the Great Panjandrum of Westminster".
25*A Specific Plan,* p. 123; *Statistical Account I,* p. cii.
26*A Specific Plan,* p. 179. Gourlay never acknowledges a debt to William Godwin, for he liked to be regarded as an innovator himself, but his emphasis on moral duty, abstract truth, evils of over-government, and the responsibility of society for its members, bears too striking a resemblance to Godwin's earlier work to be ignored.

but no bookstore would stock his pamphlet. When none would sell it, he distributed it himself. He rode out at night in all kinds of weather, for miles around, leaving copies wherever he thought they would be found—in gardens, on gateposts, in taverns, in outhouses. His sixty-nine-year-old father tried to buy up as many as possible to save his son from arrest, but he could no more deflect him from his chosen course than if he had put a pea in the path of a hippopotamus.[27]

This was the atmosphere in Fife when Gourlay moved to Wiltshire where he told himself he was going to practise his "darling pastime" of farming. The shire he intended to make his home not only contained the rotton borough of Old Sarum but also sent thirty-four representatives to the unreformed parliament of Britain. Most of them were controlled by families that dated back to the time when William I had summoned the "lande syttinge" men to meet him at Salisbury, or to the uxorious Henry VIII, who had rewarded favourites like Somerset with lands seized from the church. The most liberal of these leaders, the Earl of Radnor, made no secret of the fact that he controlled ninety-nine out of the hundred burgage tenures for the seat of Downton, one at least of which lay at the bottom of a water course. In addition, Wiltshire contained more con-centrated agricultural poverty than any other county in England.[28]

There is a saying in Fife, "Ilka doorstep has its own slippy stane." Robert Gourlay had chosen his new doorstep himself.

---

[27]M. Conolly, *Biographical Dictionary of Eminent Men of Fife.*
[28]Victoria *History of the Counties of England,* (Oxford, 1959) IV, p. 221, and p. 83, from Minutes of Evidence, Committee on Corn Laws, 1814.

7

Damning a Duke

At the same time at he was night-riding in Fife like a Paul Revere with pamphlets, Gourlay was writing south in order to clarify details in the agreement he had signed with the Duke of Somerset's steward in London on May 17, 1809.[1] He was satisfied with the twenty-one-year tenure that would give him time to reorganize Deptford Farm and repay him for the extensive modernization he planned. The rent was £600 for the first seven years and 1,143 bushels of corn (wheat) for the remaining fourteen when the revitalized land would yield much more. The duke would allow £200 for repairs to the run-down farm buildings.

He had already made plans for enlarging the farmhouse tastefully so that Jean and baby Jeany could be surrounded with the same comfort they enjoyed at Pratis, and he wished to begin the renovation immediately. He was amazed when Perrior, the present tenant, told him that in Wiltshire the incoming tenant shared the house with the outgoing till the following June. Gourlay protested that his agreement gave him occupancy of the complete messuage (house, barns and land) on October 10, 1809. Pity, said Perrior. Wiltshire custom allows me to keep the barns till my grain is sold and not to move the stock till next Lady Day.[2] Gourlay wrote to the duke to protest. The duke answered with a brief note that said it was true that the lease specified complete entry on October 10, but no one in Wiltshire ever had full entry then. Mr. Gourlay could have the usual partial entry and at the end of his twenty-one-year lease (1830) he would receive the benefit of the custom.[3] When Gourlay arrived in the south of England in September he was forced to rent a house four miles away at Stapleford, for he was a gentleman who had always maintained his own establishment.

The postponed occupancy made it difficult for him to begin his re-

---

[1] *An Address to the Special Jurymen of Wiltshire with a report of two issue trials at Salisbury, March 12, 1816*, p. 54.
[2] *Ibid.*, p. xxxii.
[3] *Ibid.*, p. lv.

organization of the farm. Deptford Farm was shaped like a tortured sausage. The house and farm buildings were situated at the narrow end that bordered on the Wily river and the tail of the land waved away up to the downs overlooking Salisbury Plain. The fields farthest from the barn were exhausted from lack of manure. An ordinary tenant farmer could not afford to pay for the labour of breaking chalk into the soil to sweeten it with slowly dissolving lime, nor could he afford to keep the four horses required to pull a dung cart through the heavy bottom land. Perrior fed twelve cows on the rich meadow grass along the river and had not fertilized the far fields.

Gourlay intended to abandon Perrior's mixed farming system in favour of a wheat-sheep economy. He would keep only four cows for domestic use, and would buy five hundred sheep. Sheep brought quick cash returns when the increase of the flock was sold, and while they were growing they served as inexpensive peripatetic fertilizing machines. A flock of five hundred pregnant ewes fresh from the meadow could fertilize a field in one night. They spent the day in the meadow too busy cropping to drop manure. At night a shepherd would herd them into a field where they chewed their cuds, emptied their systems, and with their little pointed hooves, trampled the manure into the soil at no cost to the farmer. This new "soiling system" could be put into practice only by a farmer who had enough enterprise and capital to modernize his whole system of farming.[4] By adopting this new wheat-sheep economy Gourlay could alternate crops of wheat and grass on his far fields instead of using them for pasture only. This profitable re-organization would necessitate building a barn up on the hill and a cottage for a shepherd. It would not be too costly, for Gourlay could use the wood left over from the reorganization of the home barton (farmyard and farm buildings).

The farmhouse faced the barton, a relic of ancient days when a man protected his property with the sword. Gourlay had no intention of looking out at a granary with a rotting floor and pigs rooting around a trough. He would tear down the motley assortment of buildings and erect a tidy stead-ing to one side of the house. This would provide an unspoiled view to Wily village and the downs on the far side of the Wily river.[5]

It was irritating for a man with Gourlay's energy not to be able to begin work immediately. He had come to Deptford Farm eager to impress people with his scientific farming, but every day some fresh frustration faced him. Perrior was busy selling the straw off the farm. Straw for animal

---

[4]The soiling system was introduced by John Curwen, brother of Fletcher Christian of Mutiny on the Bounty fame. *Victoria History of the Counties of England,* IV, p. 72. For other agricultural practices see Thomas Davis, *General View of the Agriculture of Wilts* (London, 1813).

[5]*An Address to the Special Jurymen,* p. 88.

bedding and for straw manure for the fields was so scarce in Wiltshire that practically every pitchforkfull was counted. Gourlay's agreement forbade him to sell the straw off the farm, but Perrior was selling his straw to the keeper of Deptford Inn across the road and was carting away load after load of dung to an undisclosed destination. In addition, his implements, which had been well cared for in Scotland, had to be left out in the weather for Perrior occupied the barns. After he cleared the clogged sluices and ditches for the water meadow, Gourlay discovered that it was Wiltshire custom for the landlord to have the ditches in order before the tenant entered.[6] All winter long letters went back and forth between Gourlay and John Gale, the duke's steward, who was impressed with his vigorous plans; Uriah Messiter of Duncanton, another steward who was not; and the duke, who did not answer because he was in London for the winter season.

His Grace was busy settling into Lord Roseberry's former house in Park Lane. His wife was "as busy as a bee upon a bank of thyme" as she furnished it fashionably with antiques. At the height of the social season in London the dung from Deptford Farm did not seem as important to them as it did to Gourlay, even though Gourlay now knew that Perrior was carting it to the duke's own tithe lands. In March the duke answered Gourlay's letters with a curt note instructing him to address all further communications to Uriah Messiter and that he was to pay all tithes, both great and small, to His Grace.[7]

The matter of the tithes from Deptford Farm was troublesome too. Gourlay's agreement made no mention of tithes, for in Scotland most tithes had been commuted to cash. Tithes in kind were anathema to the Scottish farmer. Gourlay could not find who owned the tithes from his farm. Neighbouring farms paid theirs to the lord of the manor who was the nearby Earl of Pembroke. Deptford Farm was a tithing of the parish—that is, it paid its tithes to a different owner from the lord of the manor.[8] Perrior said that the tithes belonged to him. A neighbour said they belonged to the curate. The curate said they belonged to the rector, the Reverend Mr. Dampier, who lived fifty miles away while the curate did the work.[9] Mr.

---

[6]The problem of wintering stock in the chalk counties like Wiltshire was solved by a water meadow. Grass in a low-lying meadow was kept green all winter under shallow water. In the first floods after Michaelmas sluices were opened so that a "drowner would right up a pitch" that by Lady Day would be drained off to provide a month's early feeding for stock. The meadow was then sown to hay that yielded two tons per acre.

[7]Great tithes were one-tenth of grain, hay, etc. Small tithes were butter, fowl, eggs. Scattered through the south of England were great tithe barns for their storage. Not all of the tithes collected found their way to the coffers of the church.

[8]This point was clarified by Miss Elizabeth Crittal of the Institute of Historical Research, London, England.

[9]*Apology for Scotch Farmers in England with a Case in Tythes, submitted to the consideration of the members of the Bath and West of England and the Wiltshire Agricultural Societies*, p. 17; Hammond, *The Village Labourer*, 1760-1832, p. 221; Cobbett, *Rural Rides*, 1820-1830, (ed. G. D. H. Cole, London, 1930), p. 382.

Gale, the steward, said the duke had leased the tithes from the rector. The duke did not answer. Gourlay was the new broom sweeping clean. He travelled fifty miles to see Mr. Dampier.

Mr. Dampier said he did indeed own the tithes from Deptford Farm. He would be charmed to commute the tithes to cash after a surveyor evaluated them. Though Presbyterian Robert had approached the Church of England rector with misgiving, he came away with his prejudice melted by Dampier's mild and rational religious views. Six months later, however, the duke advised Gourlay that the tithes belonged to him. Dampier had "forgotten" that he had signed an agreement with a quondam Duke of Somerset relinquishing all claim to the tithes of Deptford Farm. The duke would not commute the tithes for Gourlay but he would rent the tithe lands to him for £70 and a load of wheat straw yearly.

It was not June 25, it was July 10, 1810, when Perrior finally vacated the farm and Gourlay was free to work it. It was now too late for sowing. The fact that the scientific farmer from Scotland had no crop of wheat during his first year of occupancy was the joke of the county. Old-time residents enjoyed the discomfiture of the newcomer who had come south "to teach the moonrakers hoo to farm, mon".[10] Balked from having a wheat crop, Gourlay counted on a good crop of hay from Broadmead to feed his flock of five hundred sheep but now he found that after Perrior had moved his cows he had turned his sheep onto the river meadow. Their close-cropping little mouths had nibbled the grass to the roots. He decided to sow a late hay crop in the fields, but the couch grass choked it. In Scotland one ploughing destroyed couch grass but in Wiltshire two ploughings were necessary. Sheep would starve before they would eat coarse couch grass. Gourlay had to buy hay for his first winter's feed. He needed cash too for the rent that came due before he had drawn any income from the farm. He was chagrined to find that, although it was Wiltshire custom to allow the first rent instalment to be late, his agreement stated that his first instalment was due in March 1810, before he had even received occupancy of the farm.

Undismayed by these unexpected setbacks, Gourlay began his renovation of the farmhouse. At the back of the house were two dilapidated skillings for a dairy and a larder. He pulled them down and built a generously proportioned drawing-room, master bedroom and centre hallway. He curved a new driveway in from the road, for this was now the front of the house. He straightened the kitchen garden by exchanging two lugs of land with two cottagers. He cut down a dead apple tree and planted others. He surrounded the barn side of the property with a thatched mud wall and the house side with a four-foot brick wall topped with an ornamental iron

---

[10]*An Address to the Special Jurymen*, p. xxxii. "To teach the moonrakers" was Cobbett's comment on Scotch farmers and Wiltshire yokels, one of whom was supposed to have tried to rake in the moon reflected in a pond.

railing. He drained the road leading to the far fields and surfaced it with flints picked from the fields. He planted a thousand forest trees and many thousand quicksets.[11]

This tremendous activity around a hitherto somnolent farm indicated that the duke was getting more progress than he anticipated. He sent a third steward to investigate. This man reported that the tenant was ruining a barn by bolting a threshing machine to the floor, was stealing land from cottagers, taking flints from fields where they had lain for centuries, digging chalk where he shouldn't, pulling down buildings and converting the material to his own use, and breaking up cleve lands (pasture on the downs) that had been pasture since the time of the Romans. He reported that the duke's new tenant "was performing great waste and spoil upon the farm".[12]

In August the duke himself came to see what was going on. Gourlay was outside supervising some work. He was delighted to see the duke appear at the other side of the farmyard. He was about to stride over to give him a spirited resumé of his plans and a personally conducted tour of his improvements, but the monocle-mannered duke continued to talk to his steward without even nodding to his tenant. All Gourlay could do was glower inwardly and go into his house. He who had had a standing invitation to dine at the table of the lord lieutenant of Fife may have found it in his heart to think more kindly of the easier ways of the aristocracy of his own country than he had when he wrote his pamphlets criticizing them. The reorganization of this farm according to the latest scientific principles was just as much a creative enterprise to him as the duke's book on the properties of the ellipse. He was no ordinary farmer. His Edinburgh education was as good as any Englishman's. "Tempora mutantur et nos mutamur in illis," his pen scratched out at the duke when he recorded this incident later.[13]

The stranger from North Britain was beset by other troubles. He had not known that the rising poor rates he had undertaken to pay were £100 yearly for the farm. The first steward had told him the duke would approve £397 for suggested repairs to barns and house. Now the duke refused to authorize more than the original £200. He said that Gourlay had made the house too good for a tenant farmer and when Gourlay moved out he would be obliged to return the house to its original size. The duke reverted to asking for tithes in kind.

Gourlay now claimed compensation for the year's loss of income that he had sustained by late entry and for the amount he had expended for cleaning up the mess left by former slipshod tenants. The duke ignored his

---

11*Ibid.*, pp. 80, 101, 103, 123.
12Letter of May, 1813, *Apology for Scotch Farmers.*
13*An Address to the Special Jurymen*, p. 34.

claims for compensation, and sent his agents to collect the rent and demand entry to the house to inspect it. Mrs. Gourlay and the children were sick in bed and Gourlay refused to have them disturbed. He said that his house was his castle just as much as any Englishman's.

Again the duke's agents appeared at dawn to collect the rent. Gourlay exploded. He said he would not have his family disturbed again. He called the duke a damned scoundrel, a scandal to the land he lived in. He said he was damned if he would pay the rent till all the other matters were settled. The three agents committed his words to paper and reported to the duke that his tenant was "an abandoned and vicious character".[14]

The duke prepared a lease in place of the former agreement and threatened distraint if the rent were not paid. The new lease contained the same rent payment, but the duke demanded that Gourlay fence off twenty acres of the downs that the duke would break and cultivate, that his shepherds should occupy the two hill cottages, and that Gourlay should pasture eight of the duke's horses on Broadmead for the last year of the lease. No new land could be broken into arable fields and only one-quarter of the farm could be sown to wheat at one time. With all these prohibitions Gourlay could not employ the wheat-sheep economy that alone would reimburse him for his cash outlay, £4,000 of which he had borrowed. All the wheat he could raise on only one-quarter of the farm would be consumed in paying the rent, a system that had held more than one tenant farmer, both English and Irish, at subsistence level and prevented sufficient expenditure to maintain the fertility of the soil.

He refused to sign the lease that would keep him a dirt-scratching farmer and refused to pay the rent till he was compensated for late entry and loss of income. Early in November, 1811, the duke's agents posted notice of the sale of all Gourlay's stock and moveables. The sale realized £900, £90 more than the rent owing.[15]

Where was the universal peace that Robert and Jean had anticipated when they journeyed south? Their second daughter had been born the previous winter, and to escape the tension of the disputes over tenancy regulations, Jean had taken Jeany and Jessie in August to the sea shore at nearby Lymington. Little Jean's colour improved from the sea bathing, and when tired of walking she would run ahead of her mother calling, "Home! home to Papa!" Jean read the portions of Scott's *The Vision of Don Roderick* that her mother had copied out and sent her from Scotland, and borrowed Mrs. Brunton's *Self-Control* from the library in Salisbury. She enjoyed visits from Fife friends who did not resent her husband's political opinions. Now, as she watched their goods disappear down the road on the

14*Ibid.*, pp. xxvii, 44.
15*Ibid.*, pp. 46-51.

day of the distraint, only her pride kept her head high. All that remained
of their worldly possessions was her furniture. After the distraint there
might come notice of eviction. If they were evicted, no other owner would
rent to them. She gathered her two little girls around her for comfort and
felt her third child vibrant within her. Jeany was old enough to sense her
mother's distress. The night before, Jean wrote to her mother in Edinburgh, the child had awakened with a nightmare. Her mother ran to her
crying, "My dear Jeany, what is the matter?" The little girl sobbed out,
"Nothing with Jeany, Mama, but Jeany is sorry for Jeany's Papa!"

Jean Gourlay was worried how her husband would stand up under the
strain of these trials. She continued in her letter to her mother: "Her papa
however has been considerably better for some days past & I still continue
to be perfectly well." They had hoped that the move to a warmer climate
would benefit Robert's health. A relative from Scotland whose husband
had changed doctors wrote advising them to do the same. Mrs. Gourlay
replied that she had no faith "in the power of medicine while opposed to
the powers of discord, mischief and ill luck, . . ."

Her strong will supported her. "The kindness I feel for them [some
relatives] makes me frequently liable to forget those circumstances in my
lot, that tend to freeze all loyal hearts and prudent heads; but I can assure
you when recollection is restored to me it comes unaccompanied by any
feeling for which I can blame myself."[16] Robert Gourlay did indeed have
a good wife and he acknowledged it as he saw her "continually brighten
our clouded fate with her good sense and cheerful resignation".[17]

After the distraint Gourlay was served with a notice of eviction written
in the duke's own precise spidery script. He must vacate on Lady Day,
1812. He refused to accept the eviction. They said that he occupied the
farm under an agreement, not a lease. He said that in Scotland an agreement was as binding as a lease. Where was British justice if a duke could
scrap a written agreement? If English tenants would not stand up to their
landlords, a Scotsman would. He was no man to "bow the knee to Haman".
In England "resistance to the will of the landlord is intuitively construed
into crime," but he would take his case to the law.[18]

For a second time Gourlay's fate hung in the balance. It may be said
that he had been stupid not to inquire more fully into farming customs in
a strange locality, but undoubtedly his decision to move was precipitated
by the threat of prosecution in Fife. He was not the only man in this
situation. In the vicinity were three other Scottish farmers, "not known to

16August 11, 1811; Dec. 5, 1811, Gourlay Papers, OA. For Gourlay's health see also
    Statistical Account I, p. cxxiii. Gourlay's health will have prime bearing on the question
    of his later sanity.
17*Appeal*, p. xxv.
18*An Address to the Special Jurymen*, p. liii.

each other and invited by different landlords. . . . All were soon embroiled in trouble and ended in chancery. All were successful but all were ruined. . . . I have not heard that Lords Suffolk and Anderson ever dropped a tear over their afflictions as a mark of pity or compunction. . . . These English landlords had never known contradiction till they came into contact with tenants unaccustomed to slavish control."[19]

The "slippy stane" that Robert Gourlay had found on his doorstep was the fact that this period in history was crammed with disputes between landlord and tenant. Somerset saw no reason why he should not rid himself of an uncongenial tenant, but his tenant was not intimidated by the fact that his landlord was the second peer of the realm. He posted to London to engage Sir Samuel Romilly as his lawyer.

Romilly was no idle choice. The friend of Diderot, Mirabeau, and Bentham, in 1797 he had defended successfully against the charge of sedition a delegate of the London Corresponding Society. He supported Catholic emancipation, parliamentary and criminal law reform, and opposed suppression of public meetings and suspension of habeas corpus. In 1808 he bought a seat in parliament to put weight behind his views. Romilly assured Gourlay that he could not be evicted arbitrarily. On December 10, 1812, Gourlay's case against the Duke of Somerset was heard in the London Court of Equity and was decided in his favour.[20] Even a duke had to honour an agreement. In addition Gourlay was to be awarded damages for the careless flouting of the details of the agreement. He was entitled to compensation for loss of a year's income from the farm because of late entry. He had won his case but the catch in the verdict was that no date was set for settling the amount of the damages. Gourlay was hard up. Now the delays of the law set in, and many could be attributed to the fact that Lord Eldon, the Lord Chancellor who had heard the case, was the same Thomas Scott who had asked for the death of Thomas Hardy and Horne Tooke at their trials for treason in 1794, that he was now over seventy and the most notorious reactionary of his time. The amount of damages would not be settled until June 1816.

Jean Gourlay wrote to her mother to let her know the outcome of the trial. "My bravery, as you call it, is not assumed; but though real and lasting, it is neither blind nor devoid of sensation."[21] Now she and her husband must cultivate and keep in order the land they had been prepared to cherish as their own but which now seemed sown with dragon's teeth. Gourlay was under constant surveillance. The 1809 agreement said that he could not sell any hay off the farm. His 757 sheep had been sold the day of the distraint. He sold two tons of the 150 tons of hay he had grown to feed his

19*Ibid.*, p. iv.
20Vesey & Beames, *Reports of Cases in the High Court of Chancery*, I, p. 68; *Appeal*, p. 64.
21Dec. 5, 1811, Gourlay Papers, OA.

vanished flock and the bailiff informed the duke. Some of the remaining hay rotted and the bailiff accused him of bad husbandry. He advertised to board sheep on the farm to eat the next year's hay and was told that the land could not be let.[22]

A less determined man would have given up, but not Gourlay. He was an efficient tidy farmer who could not bear to live on a slovenly farm. He was also incapable of retiring into obscurity. When the Wiltshire Agricultural Society offered a silver cup to the cultivator of the farm that showed most improvement during 1813, he entered the contest. He knew it was not wise to spend more money while he was having a dispute with his landlord but he wrote half weakly, half magnificently, "I cannot stay my hand from improvement; it is a weakness inherent in my nature."[23] Mr. Phillips of Chadenwick Farm won the cup, but the judges said that Mr. Gourlay lost it only because he had not had time to complete his course of improvements. Not one in ten thousand, they said, could have done as much as Gourlay since he had taken over Deptford Farm.

Gourlay not only refused to retire into obscurity but he wrote to the *Salisbury and Winchester Journal* stating his side of the dispute with the duke. The paper printed his first letter and then no more. The duke's agent was allowed to answer. "The expense of litigation is to Mr. Gourlay absolute ruin—to the Duke only a drop in the ocean."[24] The duchess—who liked to be known in London as a lovely literary lady but whom Thomas Creevey, the famous diarist, called a "false devil"—wrote threatening letters to them. Once when Gourlay was absent, "She came and danced beneath my window and threatened my wife's peace of mind."[25] In spite of both duke and duchess the tenants did not move. When the local paper still refused to print Gourlay's letters, he published in 1813 another pamphlet entitled *Apology for Scotch Farmers in England with a case in Tythes, submitted to the consideration of the members of the Bath and West of England and the Wiltshire Agricultural Societies.*[26]

Some of Gourlay's Scottish neighbours had said he had opposed the Earl of Kellie out of pique. Now some of his English neighbours begged him to

---

22*Salisbury and Winchester Journal*, March 29, 1813; *An Address to the Special Jurymen*, p. 71.
23*Apology for Scotch Farmers*, p. 20.
24*Apology for Scotch Farmers*, p. 22; *Liberty of the Press Asserted in an appeal to the inhabitants of Wilts and a letter on the corn laws*, p. 5.
25*Appeal*, p. xxviii; *The Banished Briton*, #12, p. 106.
26Lieutenant-colonel F. Norman Jeans, present owner of Fisherton-de la Mere, Wylye, Warminster, Wiltshire, and Deptford Farm, talked with some elderly people who were born on the latter farm. They told him that a cousin of the Duke of Somerset had been given the farm about 1800 provided he farmed it properly and did not ask the duke for more money. The "cousin" said the house was not large enough. He fell out with the duke and emigrated to Canada. This was obviously Robert Gourlay whose impact on the locality survived a century and a half by oral tradition.

drop his suit against the duke and forget about collecting damages. He answered them, "You may be disposed to find an apology for my landlord in my peculiarities of character—but the others (meaning the three Scotch farmers facing ruin) were accommodating men. Whence come my peculiarities? Are they not bred by circumstances?"[27] There was a good deal of truth in his assertion, for the Wiltshire situation has been summed up thus: "Nineteenth century farming constantly responded to the challenge for improvement. . . . In Wiltshire, however, the results were disappointing because of the large amounts of conservatism that remained."[28]

In the dispute over Deptford Farm there were faults on both sides. Gourlay had attempted to implement with borrowed money a program of modernization that a more cautious man would have spread over a longer term. He considered that his long lease would protect him till his improvements were paid for. Time has proven that the steading he built was solid enough to stand for a century and a half and the road to the hill barn so necessary that the present owner has surfaced it with macadam. There were more issues involved than business judgement. "Gentlemen, if any of you have been accustomed to look on your tenants as vassals, such right, towards me, I will not acknowledge." His friends warned him to beware or he would render his family destitute like those of the other three Scots. "No fate which you can decree will at all influence me. . . . No, not even for the rescue of my family from poverty would I flatter your vanities or flinch from the assertion of truth."[29]

Against this background Robert and Jean gleaned as much happiness as they could from the visits of Scottish friends. One was David Wilkie of Cults, Fife, already a member of the Royal Academy. During a visit in 1814, he made a quick oil portrait of two-year-old Oliver, the heir for which his parents longed. He painted the little boy sitting like an "Age of Innocence" with a toy shovel in his arms, the image of his blonde mother. Then Wilkie's pencil sketched the parents, catching the growing delicacy that now characterized Jean's face. Small Jeany, Jessie and Oliver would soon have a baby sister Helen. Constant controversy and child-bearing were leaving their mark on their mother, but their father had barely begun to work on the second aim he had in mind when he moved south.

Late in the summer of 1813 Gourlay went to Cheltenham for his health.

27*An Address to the Jurymen of Wiltshire*, vi.
28Victoria *History of the Counties of England*, IV, p. 91. Naturally landlords had their troubles too. One lord decided to modernize cultivation by letting his land to six tenants. After one year of watching them change the century-old pattern of the fields, his wife and daughters made him buy back the leases and return the estate to its original condition. The Marquesses of Ailsbury failed because they spent too lavishly on a new manor house. The Earl of Pembroke, owner of the largest estate in Wiltshire, is said to have made the transition admirably.
29*An Address to the Special Jurymen*, p. v.

"I rode one day to the top of a hill to enjoy one of the richest, most sublime, and beautiful prospects in England. Observing some men at work in a quarry, I entered into conversation with one of them. He had 15s per week. 'You must live well then,' I said, wishing to draw forth remark, 'for the labourers in the country below have only 9s.' 'They are on the earth,' said the quarryman, 'but they do not live.' I again surveyed the luxuries of the plain, but, with the association of man pining in the midst of it, enjoyment was gone."[30]

Robert Gourlay had accomplished his reorganization of Deptford Farm. Now he was ready to begin his study of the poverty of the farm labourer in the south of England.

---

[30]*The Village System*, p. 5.

8

Tell it not in Bath!

As soon as he arrived in Wiltshire Robert Gourlay began quietly to observe
the condition of the agricultural labourers.[1] If the present provisions for
their care were to continue for many more years, he feared that "the labourers of Wiltshire would scarcely have the appearance of men: they would
be shrunk to nothing: . . . distorted—diseased—downcast".[2] In some Wiltshire parishes every labourer was on the rate. When Robert Gourlay saw
the state of Wily poor as they stood waiting for their fortnightly dole, he
suggested to artist David Wilkie that he make "Paying the Poor of Wily"
the subject of his next picture.[3]

Pasted on the front of the record book of many a harried overseer of
the poor (they were distributing eight million pounds annually) was a list
of instructions based on the recommendations of a group of Berkshire
magistrates who met at Speenhamland in 1795 in an effort to solve the
famine crisis of that year. These landlord-magistrates would not raise
wages so that the poor could buy food at inflated famine prices, for when
the famine was over it would be too difficult to lower them. They declared
that the wages of an agricultural labourer should be nine shillings per
week, and calculated that the amount necessary to keep an individual
alive for a week was one gallon loaf of bread plus threepence for extras. In
a family comprised of eight people, with a man earning nine shillings, the
wife three shillings, three children earning three, two and one shillings
and three earning nothing, the wages would total eighteen shillings. Subsistence was calculated thus: if the price of a gallon loaf of bread was three
shillings, a family of eight required twenty-four shillings to exist, plus

---

[1] During the social and economic upheaval resulting from the Industrial Revolution and
the war, certain classes prospered, namely the large landowners, the large tenant
farmers, and the industrial workers whose wages generally increased faster than the
cost of living. Two sections suffered—the handloom weavers of the north and the
agricultural workers of the south.
[2] *Statistical Account I*, p. ccclx.
[3] *Ibid.*, ccclix.

threepence each for extras. This made a total of twenty-six shillings. The parish would make up the eight shillings not covered by their combined wages.[4]

In the parish of Wily there were sixty-eight families comprised of about 350 people. Once again, as he had done in Fife, Gourlay began to talk to the people on their own level. For two years, he wrote, he "rode, walked, and watched, interfering little in parochial management". He saw eight and nine-year-old boys sent out to drive horses at plough in the severest winter weather. He saw that children seven years old were sent out to labour with no opportunity of going to school. In a neighbouring parish only three people could read and write, the squire, parish clerk, and one other. They had no way of bettering themselves. They were denied the "substantial bliss of the belly" that he considered was their birthright.

Gourlay saw Bet Bennman. Bet was poor but she was sober and industrious when she could find work, cheerful and uncomplaining when she could not. Mrs. Gourlay employed her in kitchen and sewing-room as often as possible. Somehow she managed till she was stricken "with a disease incident to women". She had no fire or bedclothes. She applied to the overseer for a blanket. First he said she could not have one. When she insisted, he tore one in two pieces and gave half to her. The blanket, curate John Ball told Gourlay, weighed twenty ounces. In February Bet asked for a doctor. The overseer said warm weather would cure her. Bet went to Gourlay for help and he procured the parish cart to take her to the magistrate in Salisbury who could admit her to the hospital. Gourlay went too, driving his gig past the flint-patterned houses and the great beautiful empty cathedral where "parson do be nattering to the cold stones". In the court house that the Earl of Radnor had just built, the magistrate refused to help and "with a face of crying misery", Bet Bennam went back to her hut and half blanket in Wily.[5]

Wiltshire landlord John Benett considered that the poor were poor because they were lazy. The poor said they had rights as well as the rich. Benett said he would grind the faces of the poor into the dirt before he would acknowledge that they had rights.[6] The poor wanted land for a garden to grow their own food. Benett said they were too stupid to garden. When the overseer tried to knock off the rate as many poor as he could,

---

[4]*Ibid.*, p. cvii. The basis of calculation, according to W. E. Tate, *The Parish Chest* (Cambridge University, 1946), p. 230, was two and a half to three loaves per week per man and one and a quarter to one and a half per dependent, plus 3d extra per man and 1d per dependent. As Gourlay's figures are usually accurate, it is evident that Wily poor were expected to exist on less than was allowed elsewhere.

[5]*Statistical Account I*, pp. cxvi-cxxi. Friedrich Engels in *The Condition of the Working Class in England, 1845*, used conditions in the southern shires to prove the necessity of radical reform.

[6]*Statistical Account I*, p. cxvi; John Benett, letter to *Salisbury Journal*, Jan. 23, 1815.

the poor said that length of residence ought to be one criterion for deciding who should receive the dole. Benett said he would pull down the cottages on his estate, and that would settle length of residence. When he was told that without cottages he would have no labourers, Benett said they could walk to work. He had many labourers who came three miles and more all winter. Their working hours were only from three in the morning till six at night and they were always punctual. Neighbouring Lord Arundel, who fed his poor indefinitely in hard times, said that Tisbury parish (John Benett's parish) was "a Parish in which the poor have been more oppressed and are in greater misery as a whole than any other parish in the Kingdom".[7] The poor called John Benett "Gallonloaf" Benett, for he said that he could cure them of asking for gallon loaves of bread as their right. Let them eat potatoes.

In 1811 Gourlay asked to be made overseer of the poor. It was not exactly a coveted position, for the poor hated the overseer if he refused them an extra shilling, and the people who paid the poor rate, based on property assessment, hated him if he raised the rate. Naturally Gourlay was resourceful enough to declare a remedy for the poverty of Wily poor. "Full employment to the poor would raise the price of labour: good wages would cause consumption: consumption would raise the price of corn: farmers would again have money: and money would briskly circulate".[8] He knew that in Fife, where the minimum wage was twelve shillings, comparatively few were destitute. He announced that twelve shillings instead of nine was the minimum wage in Wily. This was not an illegal action for either the overseer or the parish vestry could assess the parish rate. Over both the overseer and the vestry were the justices of the peace who could annul the actions of either. It did not take long for magnate-magistrate John Benett to return the minimum wage of Wily to nine shillings and remove Gourlay from his post as overseer.

In addition to offending a local magistrate, Gourlay had gone counter to the ruling economic theory of laissez faire. The hated Speenhamland system of poor relief was in accordance with the pronouncements of the great Adam Smith who said in *Inquiry into the Nature and Causes of the Wealth of Nations*, 1776, that prices and wages were determined by demand, and interference with laissez faire principles by raising wages would raise prices and demand would cease. Gourlay believed in laissez faire in trade but not in dealing with the poor. He still believed in the statement he had made in *A Specific Plan for Organizing the People Independent of Parliament—To the People of Fife . . . of Britain!* that he was an agent of a society that ought to look after its unfortunates and reduce their number.

[7]Hammond, *The Village Labourer,* 1760-1832, p. 267.
[8]*Statistical Account I,* pp. clxx, cix; *The Village System,* p. 26.

But parliamentarians in their speeches took their examples from the past world of Cato and Cicero rather than from the times in which they lived. If they adopted a theory from the new science of political economy, it would come from the books of such political pessimists as Malthus and Ricardo who drew their conclusions from things as they were, not from things as they ought to be, or from Adam Smith whose advice was not to interfere.

During their brief reign in 1807 the Whigs had introduced a new Poor Law, but Parson Malthus told Whig Samuel Whitbread (called Fermentation Sam because of his brewery) that if the parishes built cottages for low rental it would encourage procreation. When the Tories returned to power, they said they could do nothing for the poor till the war was over. The war was far from being finished. Napoleon would soon name his son King of Rome and prepare to march on Moscow.

When no action was forthcoming, extremists rose. The Luddites, tired of waiting for parliament to listen to their needs, began to smash the machinery they said was stealing their work. Parliament could not find their leader, although the Green Bag committee sat daily to investigate every clue.[9] Maurice Margarot, who had been deported to Australia in 1794 with Thomas Muir, had returned and was seen in the company of some Fife weavers who wanted higher wages.[10] The greatest danger of all had come in May 1812. While Henry Brougham was urging the repeal of the Orders-in-Council that were the cause of part of the distress, the sound of a shot came from the parliamentary lobby. Prime Minister Perceval had just been assassinated by a man who thought parliament talked too much and did too little. William Cobbett, who had been released from prison where he had been held without trial, wrote openly in the *Political Register* that the only reason the government continued the war was because it would have to reform parliament if the war were concluded.

Reform was far away. The ruler of England was now the Prince Regent, for George III had lapsed into permanent insanity in March 1811, when his favourite daughter, Amelia, died. At the time of the Prince Regent's first levée, the counties met to vote addresses of loyalty. A laudatory one came from Fife where Kellie was preparing to write to Hawkesbury: "I am now extremely sorry to mention to your Lordship that from Lord Caithness' very dangerous state of health I fear we will soon lose that worthy Peer. In that event I beg leave to solicit your Lordship's interest for the appointment of Postmaster General for Scotland . . ."[11] When Lord Morton followed Lord Caithness' example, Kellie indicated he would enjoy being

---

9All reports of violence, such as women seizing flour from ships to prevent its export, were put in a green baize bag. See William Cobbett, *The Luddites, or the History of the Sealed Bag.*

10H. Meikle, *Scotland and the French Revolution*, p. 219.

11Kellie Castle, Oct. 28, 1813, Liverpool Letter Book, British Museum.

lord lieutenant also. Fife would send laudatory addresses as long as Kellie lived. The English town of Poole was bolder than Fife. It saw no reason to felicitate the Regent when it contemplated the state of the nation, "the disastrous . . . climax of exhaustion at which our finances have arrived" and "the system of prodigality which has so long fed upon the vitals of our country".[12]

A woman who stole potatoes from a cart in Manchester was arrested for treason. When the Luddite leaders were arrested, reformer Sir Francis Burdett claimed that starvation, not sedition, made them break machines. Lord Byron limped into the House of Lords to make his maiden speech in defence of the people whom the government was proposing to hang. Whig Samuel Whitbread said that the condition of the English village poor was worse than that of the negro slaves in the West Indies. At the end of June 1812, the Prince Regent sank his growing succession of chins into the great cravat he wore to hide them, and sent a message to parliament ordering it to stop the rioting in the country.

The five royal dukes (who had voted against the abolition of the slave trade[13]) were silent. Whigs Grenville and Grey were silent. William Cobbett began to write, "Watch the Whigs", for many of them feared reform. Robert Gourlay wrote that peaceful petitioning should be the people's method, not burning and smashing. Parliament counselled patience till the war was over.

Suddenly, the war did end. On April 11, 1814, Napoleon abdicated his imperial throne. All over Britain oxen were roasted in market places and candles of gladness flickered in every window. The Duke of Somerset wrote to his brother that the balance of power must be restored immediately. Prince Metternich came to London to confer about turning the clock back to 1789, to restore the toppled crowns of Europe in the interests of the balance of power that had not prevented the war in the first place. As Louis XVIII proceeded to Dover to embark for France, two English princes escorted his carriage. Robert Gourlay sent to the peace celebrations as much produce as he could spare from Deptford Farm with the comment that peace without reform was not sufficient.

Peace had come to a country that had been twenty-two years at war. For the first time in a quarter of a century parliament could not tell the people, "Wait till the war is over". Instead they were informed that the hardships of the post-war depression that set in were caused by the transition from war to peace.

Gourlay felt the pinch of transition from war to peace and reasoned

---

[12]*Salisbury and Winchester Journal*, March 4, 1811.
[13]Sir Samuel Romilly, *Memoirs* (London, 1841) II, p. 11. Nathan Rothschild, leaning against "Rothschild's pillar", made a killing on the London stock market when he received news of Waterloo by special courier. A money-making brain never really slumbers.

that he should plant less wheat and more hay in the spring. He also had a few ideas about increasing agricultural production in Wiltshire. He had joined two newly formed agricultural societies, the Bath and the Wiltshire and South of England societies. When he found that they were not as progressive as he hoped, he became openly critical. They had offered a prize for the best woman plougher! That was an insult to womankind, he said. Wiltshire farmers ploughed with a man at the plough and one at the horse's head. One man was enough, proved Gourlay whose ploughmen won prizes at the local fairs. Some Wiltshire farmers sowed turnip seed broadcast as if they lived in Bible times. They should drill straight rows for seeding, he said, and carried off the prize for Swedish turnips. Gourlay paid a living wage so that his workers worked "without scold or scowl". Tension was building up till finally Gourlay (one of nine vice-presidents) declared that agricultural societies were vain and trifling.

On November 21, 1814, he published *Address to the Farmers of the Hill Country of Wilts, Hants, and Dorset* in which he announced three contests of his own that were designed to amend three faults of agricultural practice in Wiltshire.[14] First, he challenged them to plough without using children, for he considered inhuman the practice of refusing parish aid to families who would not send their seven-year-old children into the fields even in winter. Second, he offered a twenty guinea prize for ploughing with one driver, and for growing turnips by drilling. Lastly, he offered to be one of twenty farmers to put up twenty guineas yearly for five years for the farm showing the most improvement. He considered that a cup was not sufficient incentive for the expenditure involved in improvements.

The placards announcing this had barely been nailed in the nearby market places before he posted new ones announcing a bold political challenge entitled "Institution for the Benefit and Protection of the Farming Interest". It was a call to all tenant farmers who farmed land over the value of £100 yearly to sign a paper that he would leave at the Crown and Anchor in London signifying their willingness to attend meetings to elect deputies to meet at seventeen designated centres throughout Britain. There they would elect seventeen representatives to meet him in London to discuss the interests of tenant farmers as opposed to the landed interests. He said, "The interest of the great landlord had ever been too remote for delicate feeling; hence he suffered the Work Horse Tax, the Sugar Distillation, and the Farmers' Income Tax, of all the impositions the most unfair and degrading. Blessed, indeed, shall be our present necessities, if they rout from its strong hold the last cohort of feudal power".[15]

14*Statistical Account II*, p. vii.
15*Statistical Account I*, ccxxvii. The taxes he mentions bore more heavily on the tenant farmers than the landowners, just as the tax on beer penalized the man who bought his pint in a pub, not the rich who brewed their own.

It was such bold talk that Gourlay was afraid to sign his name. He merely placed at the end an acrostic-like assemblage of letters: R G F D W F O P F. No farmers signed the paper requesting them to come to London "sufficiently zealous to take the lead in trouble". What Gourlay was asking for was the equivalent of an agricultural corresponding society, and political corresponding societies had been outlawed twenty years before.

It was still not Gourlay's nature to hide behind anonymity. A month later he revealed that the initials stood for Robert Gourlay, Farmer at Deptford, Wilts, formerly of Pratis, Fife. This caused his expulsion from the Bath and Wiltshire Agricultural Societies, though the reason they gave was that he was engaged in a lawsuit with his landlord who was a patron of one society. He would write a few years later that he did not care what the "old women" of the Bath Society did, that he was "as proud of this record as St. Paul was of his dozen odd whippings", but he did care, for he was sensitive about the allegation that he was peculiar.[16] No reformer can afford to be sensitive. Cobbett thrived on opposition, and even enjoyed using some unreformed methods as he battered his way to the polls. Gourlay's nervous reaction to his uncertain position caused part of the ill health that his wife acknowledged plagued him intermittently.

Now another blow was dealt him. Because of the reports that were reaching him from Wiltshire of his son's activities, his father disinherited him.[17] Oliver had already expressed disapproval of Robert's political statements by stopping payment of his son's annuity as specified in his marriage contract, but that was understandable for Robert owed him money. Disinheritance was more than an embarrassment for both had signed for a loan from Jean Gourlay's brother, Thomas Henderson, on a heritable bond. This light-hearted borrowing back and forth had been a trifle in the wartime boom when Oliver was worth £120,000 with £20,000 in floating capital. Now every lender and banker was calling in money. The note was binding on Robert if his aging father died.

In spite of the precarious condition of his personal affairs, Gourlay did not ignore public matters. He began to attend the political meetings that were becoming more violent in the southern counties. Reformers Cobbett and Hunt contested as many elections as they could in their endeavours to add to the two reform seats in Westminster. Gourlay went to one county meeting to express his regret for Cobbett's recent imprisonment for his fearless articles in the *Political Register*. Cobbett shook Gourlay's hand warmly. Gourlay came away convinced that Cobbett was a man whose

---

[16]*Statistical Account II*, p. xvi; *Canada and Corn Laws; or No Corn Laws, No Canada*, p. 12.
[17]*Statistical Account II*, pp. vi, xiv; *The Banished Briton*, #25, p. 307.

name would be as lasting as England's history.[18] Nevertheless he could not endorse Cobbett's methods of bringing about parliamentary reform, for he seemed to attract a maelstrom wherever he went.

In 1815 the parliament they were both working to reform passed a law legalizing the use of iron man-traps to catch poachers in milord's woods. The issue of poaching was not all one-sided, but the law indicated that legislators and people were enemies. Gourlay still believed that the poor were pawns caught by economic forces too strong for them to control without help. Since coming to Wiltshire, he had worked mainly for himself. He was now ready to champion others.

---

[18]*Statistical Account I*, p. ccxix.

9

Alas! Luckless Petition!

First Gourlay took up his pen in a war against tithes, one of the systems that prevented England from feeding her hungry people. On February 24, 1915, he published *The Right to Church Property secured and the Commutation of tithes vindicated, in a letter to Rev. William Coxe.* It was his reasoned contribution to a sequence of pamphlets on the subject after John Benett had won the 1814 Duke of Bedford gold medal with an essay advocating commutation of tithes by purchasing land for the clergy. He was answered by Archdeacon Coxe who claimed that tithes had been ordained by God. Gourlay denied the sacrosanctity of tithes. Any system was bad, he concluded, that extracted ten per cent of farm produce from manure to vetches, no matter how great was the need of the farm to retain them for the good of the soil.

Next he entered the current controversy over the Corn Laws. Gallon-loaf Benett was galloping around the county with a corn bill petition hanging half out of his pocket to secure support for a law to raise the duty on corn. He called a meeting at Warminster, January 6. Henry Hunt rose to inquire why no tradesmen were invited to the meeting, for any rise in the cost of living was their concern. Benett's thoughts did not run in a democratic groove. He announced that only those who agreed to the proposed law were welcome to his meeting. Those clamouring outside the locked door would be intruders if allowed to enter. A vote to open the meeting was called and won. They adjourned for more room to the Town Hall and then to the Lord's Arms Inn. There Benett was forced to listen to his opposition tell him that landholders lived in luxury by *grinding the face of the poor,* and that the country had had enough of quacks and agricultural societies.[1]

---

[1] J. Benett, "Evidence given before the committee of the House of Lords on the Corn Bill", January 9, 1815; Freeholder, *A Word at Parting, being Algernon's address to the freeholders of Wilts stating some . . . of the occurrences at the late contested election for the county.*

Gourlay claimed that the landowners kept the price of grain high so that they could charge high rents and report that the country was prosperous. His recommendation was an ad valorem duty on corn soon to be abolished in favour of free trade. His other suggestions for the revival of the country's economy and lessening of poverty were speedy improvement of agricultural methods and taxes on interest derived from idle capital, but no taxation of industry that was still so young in Britain that it needed all its capital for development. Because Britain's main riches lay in land, a well-administered property tax ought to supply the bulk of its revenue. Instead, the present land tax produced half a million less revenue than it had a century before and the taxes on consumption had increased by thirteen million. When the *Salisbury Journal* refused to print his views, he published another pamphlet, *Liberty of the Press asserted in an appeal to the inhabitants of Wilts and a letter on the corn laws.*[2]

Ten days later, he set off for Bath with another manuscript in his pocket. The typesetter was drunk, but on March 13, 1815, *Tyranny of the Poor Laws exemplified* was published. It described poor Bet Bennam's struggle for decent existence. "Englishmen!" Gourlay appealed, "You do not know the iniquity which has reduced the kingdom to pauperism; . . . It is not the laws so much as arbitrary power which you have permitted to grow up and overshadow the views of benevolence, and the substantial ends of justice."[3]

When the first edition of *Tyranny of the Poor Laws Exemplified* sold out, he ordered a second to send to every member of parliament. To this he added a letter asking the people of Wiltshire to give him their confidence so that he could reform the system and eradicate poverty. Some people were rioting and breaking machines. Violence was wrong. The people must petition. Some said parliament ignored petitions, claiming the signatures were falsified. At a recent political meeting tables had been set up for people to sign a petition but many had to leave before their turn came.

Gourlay proposed a foolproof plan by which the whole nation could petition by parishes in which every one knew each other. All the petitions would be gathered into a great book for presentation to parliament. He prepared a petition asking for a law that "from Michaelmas to Lady-day, no overseer, or any other person, shall have the power to call out children, under twelve years of age, to labour, or to withdraw any stated allowance from the parents or guardians of such children, because of their not labouring; at least, if such parents or guardians do put such children to school during the said period of each year." This was signed by nearly one hun-

---

2*Liberty of the Press*, p. 14; *Statistical Account I*, pp. cxlviii, vi, ccclxiii.
3*Tyranny of the Poor Laws*, p. 8; *Statistical Account I*, p. cxx.

dred members of Wily parish, some with an X.[4] At his own expense Gourlay mailed seven hundred copies of *Tyranny of the Poor Laws exemplified.* His only acknowledgments came from the two Whigs, Lord King and Francis Horner, one of the founders of the liberal *Edinburgh Review.*

Gourlay approached Mr. Paul Methuen, to present the petition. Methuen was a new type of Wiltshire member of parliament, the first to break down the long monopoly of power held by four leading families, the Seymours, Longs, Herberts and the bishops of Winchester. Methuen was backed by one of the newly formed political clubs.[5] Now, though comparatively naked of cognomen, he represented the county of Wiltshire along with William Pole-Tylney-Long-Wellesley, nephew of the Duke of Wellington. Methuen presented to the House of Commons Gourlay's petition to raise the working age of farm children. Lord King presented it to the House of Lords. Both were tabled. It was a disappointing conclusion to Gourlay's maiden effort at influencing parliament, but it indicated that he had left county politics behind and was now aiming at the national level.

Political meetings in England were becoming so violent that candidates now must pay for their own police supervision at the polls. Reformers Cobbett and Hunt were more active than ever in their efforts to be elected on a platform of peace, no graft, electoral reform and abolition of the Corn Law of 1815 that parliament had passed in spite of opposition. Cobbett and "Orator" Hunt had temporarily joined forces, a great sacrifice for them to make, for the reformers enjoyed vilifying each other as much as they did their enemies. Gourlay did not join them because of his disapproval of their stand on the property tax and he said so at a political meeting. This was one of the rare occasions on which he spoke. He found that the depth of his feelings on such occasions caused his words to tumble out so precipitously that he became almost incoherent. This he termed "a natural frailty".[6] He often passed out handbills presenting his views rather than trust himself to oration.

Once again he was busy with his case against the Duke of Somerset. On January 9, 1816,[7] he was offered a lease in place of the original agreement, but he refused to accept it. Its stringent terms demanded £40 extra rent for every acre of downs broken up and £20 for any acre of meadow

---

[4]*Ibid.*, pp. cxxii, cxxxvi, cxxxix. The Factory Act of 1819 prohibited employment of children under nine—in cotton factories only, and with no system of inspection.
[5]Victoria *History of the Counties of England,* V, p. 204.
[6]*Letter to the Earl of Kellie,* p. 45.
[7]The income tax that had had such a fatal effect on his fortunes in Fife was rescinded this year. The repeal of the income tax was the first Whig defeat of the government in years. It was a personal triumph for Brougham, but it was not in the best interests of the county. A factory owner with no land but a £10,000 income paid less than a farmer who might gross £500. A. Aspinall, *The Passing of the Whigs,* (University of Manchester, 1947).

mowed twice a year. In March, four years after the London court had ruled that the Salisbury court was to determine the amount of the damages due to Gourlay, his case finally came before a special jury.

The courtroom at Salisbury was filled. Gourlay was defended by Solicitor-general Sir Robert Gifford, and employed a shorthand reporter to record the trial. When Perrior was called as one of the duke's witnesses, he claimed that he had given Gourlay possession of Broadmead on April 23, 1810. Murmurs of dissent were heard. The duke's main charge against Gourlay was that he had made Deptford farmhouse too good for a tenant farmer. Gourlay was awarded £1,325 damages plus £625 subject to the ruling of the Master of the Rolls. The Master of the Rolls was also to settle the terms of the lease, not the duke's agent. The catch in the verdict was that none of the larger sum was to be paid till the £625 sum was settled. Gourlay published *Address to the Jurymen of Wilts with a report of two issue trials at Salisbury*, March 12 1816, the book that would fulfil ironically another of his many ambitions: "I had the idea to publish the minute rise and progress of a lawsuit three years ago in a small pamphlet".[8]

On April 2, 1816, he published another pamphlet, *To the Labouring Poor of Wily Parish*. It was his answer to a recent statement by the chancellor of the exchequer that "the independence of the poor must be the result of their own exertion"; to a statement by Mr. Curwen that the poor should give four pence out of every shilling they earned for the relief of the farmers who paid the poor rate; and to a statement by the secretary of state who said that the poor had no right to petition. Again he urged them not to burn ricks like "Swing", or break machines like the Luddites. Peaceful petitioning would lead the people into their rightful power. Then his Presbyterian upbringing prompted him to add: "I point to those divine lessons of morality which shine from the Christian code . . . which are a lamp to the feet and a light to the path of all from beggar to king. In this clear light I wish the poor man to walk boldly, asserting his civil rights and seeking for himself and his fellow labourers a reasonable share of the good things of life—independence, not wealth, liberty, not licentiousness."[9]

If ever it was hard for the poor to listen to reason, it was during this cold wet spring of 1816 when there was no prospect of a good crop. As Gourlay travelled to London in August 1816 to defer payment of a £400 note, he saw that even the potato crop was doomed to failure. Around London the fields were flooded by the swollen Thames. This winter the poor would be cold as well as hungry for it was too wet to cut turf for fuel. To consider the crisis, parliament called a meeting of dukes, churchmen, economists, and even Robert Owen who had influential friends in spite

---

8*An Address to the Special Jurymen of Wiltshire*, pp. 7, lv; *Appeal*, pp. 64, 65, 76; Vesey, *Reports of Cases in Chancery*, XIX, pp. 429, 432; filing of case in 1812, I V&B, p. 68.
9*To the Labouring Poor of Wily Parish*, p. 17.

of his radicalism. In 1813 he had published *New View of Society* that contained the germ of socialism. Owen was asked to prepare a report for the committee. He reported that the only way to keep the poor eating was to provide employment for them in Villages of Cooperation. Brougham, stunned by this radical recommendation, refused to present Owen's report to the committee. Owen was recommending economic equality to a country that had not yet achieved political equality.[10] Daily the poor grew hungrier, but there were iron man-traps in the fields where the great hares lollopped in plain sight, and a man could be sentenced to seven years' deportation for being caught in the woods at night with a net or trap.

The national situation was desperate but Gourlay himself discovered a bright spot during his latest visit to London. While he waited to see Sir Samuel Romilly about speeding the settlement of his £1325 and £625 payments, he breakfasted with his cousin Alexander Wilson. Wilson had recently visited Deptford Farm in such a state of gloom that he was ready to commit suicide. The House of Anderson had failed in 1811 when Wilson's partner had contracted debts without his knowledge. This morning Wilson was bursting with the news that he had just won a lottery prize of £40,000. Was it honest of him to keep this money or must he use it to pay the debts contracted by his partner? Cousin Gourlay's opinion was that the money belonged to Wilson since he had not personally contracted the debts of the House of Anderson.[11]

After this happy meal, Gourlay tried to defer payment of his note for £400, but was refused. Remembering the money that was burning a hole in Wilson's pocket, he sent a messenger to him asking for a loan. Wilson obliged by paying his cousin's debt of £429 8s 10d. The Duke of Somerset, in London for the winter season, heard that his troublesome tenant had discharged a large debt. Though Somerset owed his tenant five times as much as his tenant owed him, he despatched an agent to demand half a year's rent. Grandiloquently, Gourlay borrowed the rent too from the cash-happy Wilson. He now owed Wilson £739 14s 9d. Bolstered by the thought of his cousin's secret affluence and the hope of speedy payment of the money the duke owed him, he gave another creditor, Mr. Thomson, a note for payment of £900 in three months.[12]

Soon after he returned to Deptford Farm, he thought of a scheme for disposing of the four years' accumulation of hay that he was forbidden to sell off the farm. If he could borrow more money from Wilson, he would buy another flock of sheep. He returned to London, but Wilson's cir-

---

[10]G. D. H. Cole, *Life of Robert Owen* (MacMillan, 1930), p. 175.

[11]*The Banished Briton*, #25, pp. 302-324, 326-7.

[12]Was Gourlay prodigal? See K. Feiling, *A History of England*, p. 805: "All the new corn-lands, often enclosed on borrowed money, threatened to be a dead loss."

cumstances had changed. The House of Anderson was being sued for £20,000 which it had lent to another firm that was now facing bankruptcy. Gourlay returned home with his financial problem unsolved. He could not even seek consolation from his wife, who never failed to comfort him, for he had sworn to keep Wilson's secret.

In December 1816, the Master of the Rolls disallowed the recommendation of the special jury in the spring that an additional £625 was due Gourlay. He could now hope for receipt of the £1,325, but he knew that the duke would delay paying it.

As the year 1816 drew to its wet starving close, any person with a solution for the national need could gain a following. For some years, a former teacher named Thomas Spence had been walking through Westminster alleys passing out buttons bearing the words "Spence's Plan". Spence's Plan for eliminating poverty was land nationalization—give the land back to the people from whom it had been stolen in years past and let the farmers pay their rent into a common fund for the benefit of all. Spence had recently died but the people who still wore his buttons were preparing to hold a meeting in Spa Fields, just outside London, to discuss a petition to the Prince Regent himself. In December they invited Orator Hunt to address them. The magistrates sent local yeomanry to oversee the gathering. When they overheard the words, "Spence's Plan", they thought a plot had been hatched.[13] Violence ensued. The papers blazoned next day that the country had narrowly escaped revolution. They said that printer Beckwith's shop had been sacked and a Mr. Platt fired at and nearly killed.[14]

Gourlay was in London again to see his lawyer. He went to the Spa Fields meeting to confirm his opinion that petitioning by parishes, not by unwieldy meetings, was the best way to gain reform. He knew therefore that the actual meeting had been orderly and that the trouble had been caused when a splinter group had left the area to parade through the streets. He also knew from his reading of history that mock plots are often part of government policy to awe public opinion into endorsing repressive methods. He doubted the newspaper reports of the violence. He went to Beckwith's shop to interrogate the man who said he had seen Platt fired at, and the man could not show him a single drop of blood on the floor where Platt was reported to have lain, mortally wounded.

Henry Addington, now Lord Sidmouth, head of the Home Office, and Lord Castlereagh, now foreign secretary and leader of the House of Com-

[13]G. Wallas, *Life of Francis Place* (Knopf, 1918), p. 61; Place Papers, British Museum, Reform Volume, 1811-1819; *The Banished Briton*, #16, p. 159; Gourlay's opinion of Spence, *The Petition for the Benefit of the Labouring Poor*, p. 32.
[14]People paid a fee to read in shops like Beckwith's.

mons, believed the reports.[15] They announced that the depravity of the people warranted suspension of the Habeas Corpus Act, its only peace time suspension in history. [16] The government began to prepare a series of other "Gagging Acts". There was death for any citizen who prevented soldiers from suppressing bread riots or dispersing meetings for electoral reform. A magistrate could command any meeting to disperse and any one who resisted him could be hanged. All public reading rooms such as Beckwith's were closed. As soon as the Gagging Acts were passed, the stock market rose, for the leaders of the country now felt secure from the sansculottes of the furrows and the shuttles.

On January 28, 1817, the Prince Regent was hissed and his carriage stoned as he returned from opening parliament. Once again the lord lieutenants collected signatures for loyal addresses. One lord lieutenant resigned for he could no longer serve a government that passed Gagging Acts. The man who was regarded as a monster in Fife and a peculiarity in Wiltshire began to compose another pamphlet that he published on January 1, 1817, under the title of *The Village System, being a scheme for the gradual abolition of pauperism, and the immediate employment and provisioning of the people.* It was an expansion of the dream he had had when he finished his tour of England in 1801. It was dangerously close to Spence's Plan for land nationalization and Robert Owen's Villages of Cooperation.

Gourlay wrote that if the poor had half an acre of land per family, they could be eating by the first of July.[16] The government should expropriate one hundred acres in each parish. Landowners would be compensated for the land which would be rented to small holders who would be styled freemen as soon as they had been off the rate for two years. Robert Gourlay was tired of patches of improvement like his own little efforts at Pratis and Deptford, like those of Lord Estcourt in Wiltshire, like Lord Belton's in Rutland, like poet Robert Southey's and his scheme for a pantisocracy on the banks of the Susquehanna, like Robert Owen's at New Lanark. There had to be a national plan if poverty were to be eliminated.

The *Salisbury Journal* would not even advertise Gourlay's pamphlet, *The Village System,* but it printed for free distribution a letter from "A Villager" entitled "Mr. Robert Gourlay, whipped with twigs from his own birch". "Mr. Gourlay, sir, is a revolutionist. A bold and decided one. Never,

---

[15]Addington, son of the doctor who attended Pitt's father during his mental breakdown, is rarely characterized as being anything but mediocre. He failed to realize that the years 1816-1819 constituted a terrible peak of poverty.

[16]At this time the parishes were spending on poor relief an amount equal to one-fifth of the national budget, and all that the expenditure achieved was the anger of both rich and poor. S. and B. Webb, *English Local Government from the Revolution to the Municipal Corporations Act* (Longmans, 1906) I, p. 4.

since the days of Paine did Pamphleteer more nakedly expose himself or his motive. . . ." Mr. Gourlay, said "A Villager", was dominated by pride, conceit, sourness and dogmatism instead of the sweetness and docility he claimed. He owed his present misfortune to his own cold, stern, litigious and selfish pride.[17]

There was truth in this assessment of his character, but Gourlay was too busy to bother with "A Villager". He had drawn up a petition setting out in detail his scheme for land for the poor and had it signed by ninety-five members of Wily parish. He asked the two county members, Mr. Methuen and Mr. Long, to present it to parliament. As they were not going immediately to London, Gourlay took the petition to London himself. Lord Cochrane said he might present it if the question of the poor were not a delicate one. Burdett said he would present it if it were copied on parchment. Henry Brougham, aspiring to be leader of the Whigs, told Gourlay he would present it if Sir Francis did not. Sir Francis accepted it after it was copied on parchment.[18]

On the evening of February 10, Gourlay was in the gallery of the House of Commons to await the presentation of his petition. Sir Francis, top-hatted, top-coated, strap-trousered, impressive, was in his seat reading Gourlay's petition by the light of the candle on his desk. He read it, re-read it, showed it to members around him, put it in his pocket and left the chamber. Two days later he returned the petition saying he had not had time to read it. Gourlay took his petition to Lord Cochrane. Lord Cochrane was at Sir Francis' house and read it when he returned. "But", Cochrane said fatuously, "if every man had a garden, how would the gardeners make a living?"

On Sunday Gourlay went to Major Cartwright's house to compare their proposals for parliamentary reform. Their only disagreement was on the matter of annual versus triennial parliaments. Lord Cochrane came in, and seeing Gourlay, returned his petition saying that he did not want to complicate the fight for the reform of parliament with a fight for the poor. Gourlay took it to Methuen, who said the language was too weak, and then to Lord Folkestone, who said the language was too strong. "Here", said Lord Folkestone, "take your Petition. I cannot present it: it is not a Petition, it is an act of legislation." Alas, thought Gourlay as he left Lord Folkestone's house, "Alas! luckless Petition! thou art either too bad, or too good for the Lords of this generation."[19] A few days later Methuen presented the petition. What was the explosive result? Nothing. It was tabled.

On March 13, 1817, Gourlay published his petition and the tale of his efforts to get it presented in a pamphlet with the resounding title, *Petition*

17*Statistical Account II*, p. xiv.
18*Petition for the Benefit of the Labouring Poor*, pp. 20-25.
19*Ibid.*, p. 26.

for the benefit of the labouring poor, presented and not presented by Sir Francis Burdett, discarded by Lord Cochrane, spurned by Lord Folkestone, now laid before parliament, with occasional correspondence and remarks on the subject of the poor laws, parliamentary reform, etc.

The day after the publication of this pamphlet, Henry Brougham rose in the House to give what has been called his great speech "On the State of the Nation". With his bulbous nose twitching, he urged parliament to recognize that the crisis was real and the Gagging Acts were not the proper treatment for it. "Mistaking the symptoms for the malady, we have attempted to stifle the cries of the people in their extreme distress, instead of seeking the cause of their sufferings and endeavouring to apply a cure." Brougham moved that the House go into committee to consider the state of the nation. His motion was lost, 118 to 63, and the Royal Speech at the close of parliament blamed the situation on the transition from war to peace.[20]

Gourlay, whose aim was to sting till the blood ran, had written at the end of his pamphlet: "A committee sat on the subject of the poor laws last summer and . . . did nothing: a committee is sitting now upon it but they will sit till their eggs are addled under them unless they adopt liberal principles". Not even a gadfly could pierce the elephantine hide of the unreformed parliament. Former famine years in 1795, 1800, and 1812 were like prosperity compared to 1816-17. In Birmingham alone 27,500 of a population of 84,000 were on parish relief. "I knew you to be a man of great income," Gourlay wrote about rich Sir Francis Burdett who had married the even richer daughter of banker Coutts, "but the grand delusion of our present system is to keep one part of society down, that the other may revel in superfluity."[21]

Burdett answered that the writer of petitions from Wily was a deluder of the people himself, for he made a set of ignorant people the tools of his own restless pride. Another critic, MM, wrote in the same issue of the *Salisbury Gazette*, April 10, 1817, that Lord Folkestone was right to sent Gourlay packing. If the poor were given a slice of someone's land, next they would demand a slice of his house. The arguments were as old as time, but this year's season for planting was slipping away, and the poor could not eat the daffodils.

If none of Gourlay's efforts seemed effective, what were other reformers achieving? Burdett and Cochrane considered electoral reform the panacea for the ills of the country. So did William Cobbett who begrudged every emigrant who left Britain, for it meant less support for his campaign for parliamentary reform. Cobbett scorned Robert Owen's Villages of Co-

---

20*Ibid.*, p. 4; *Statistical Account I*, p. cxli, from Feb. 5, 1817, petition.
21*Petition for the Benefit of the Labouring poor*, pp. 37, 41, 42.

operation, calling them Parallelograms of Paupers.[22] When Owen's socialist experiment failed, he began to concentrate on a bill to humanize factory working hours. On the days when the papers mentioned his campaign, he bought 30,000 copies to give away. The mail coaches were often late on the days when Owen despatched his papers. On one occasion he had so many petitions to present to parliament that he hired a coach to carry them.[23] One was signed by 40,000 people. The result of two years of Owen's constant lobbying was a Factory Act that applied only to cotton factories.

Compared with Owen's Niagara of literature, the seven hundred copies of a poor law pamphlet that Gourlay sent out were only a trickle. Owen owned a lucrative factory, while Gourlay was a tenant farmer engaged in a dispute with his landlord. He did what he could. His first poor law pamphlet circulated as far north as Glasgow. A friend in that city wrote him that he had read it and found its recommendations similar to those of a Mr. Owen of nearby Lanark Mills. Gourlay obtained a copy of Owen's *Essays*. He acknowledged the similarity of their views but said that, since he had described his village organization in 1800, he should have the credit for the originality of the scheme.[24] A friend of Edward Wakefield wrote Gourlay to ask for a dozen copies of the pamphlet for Wakefield to give to his friends.[25] That is all that is known of the circulation of Gourlay's pamphlets except that two of them, *Apology for Scotch Farmers* and *To the Labouring poor of Wily Parish*, reached second editions, and not one went unnoticed.

The eighteenth century has been characterized as a brilliant but heartless age. The nineteenth had barely begun to benefit from its brilliance and recover from its heartlessness. It was a highly religious age, but Owen and Cobbett turned atheist because of the shortcomings of current religious teachings. Gourlay used the Bible to point out his path, but not like many divines who taught that poverty was the result of immorality, laziness or God's will. His friend Thomas Chalmers praised the virtue of charity. The Scot had just been chosen to preach the anniversary sermon of the London Missionary Society. The church was filled four hours before "the tartan" rose to expound his text. George Canning, secretary of state for foreign affairs in Lord Liverpool's cabinet, was among those who listened.[26] Few heeded the words of wicked and poetic Lord Byron who had told the House

---

22G. D. H. Cole, *Life of William Cobbett* (Collins, 1924), p. 194, quoting from the *Political Register*, June 4, 1814.

23G. D. H. Cole, *Life of Robert Owen* (MacMillan, 1930), p. 188.

24*Statistical Account I*, pp. cxxxviii, clxi. Godwin's *Political Justice*, 1793, presaged a society of loosely linked communities in which the rule of reason gave each man his needs. Spence's pamphlets, about 1800.

25*The Village System*, p. 33.

26M. O. Oliphant, *Thomas Chalmers* (Methuen, 1905), p. 54. The text was I Cor.: 14.

of Lords that never, never in any heathen country had he seen such misery as existed in his own so-called Christian country.[27]

Now Gourlay had to shelve efforts to relieve the suffering of others, for he had his own family to feed. Another daughter, Catherine, had been born to him and Jean on March 25, 1815. Another called Barbara, "to whom the niggard church denied its prayers", died before she was baptized. The oldest in his family of five was nine. Now he began to think in earnest of the land he and his wife possessed in Canada. Twice they had been visited at Deptford Farm by cousins (once removed) of Mrs. Gourlay's from Canada. In the summer of 1810 the Honourable William Dickson, member of the legislative council of Upper Canada, had come to consult with Mrs. Gourlay's mother (Jean Hamilton Henderson) about the choice of a Scottish school for her brother Robert Hamilton's sons by his first wife.[28] The Honourable Mr. Dickson was executor of the Honourable Robert Hamilton's considerable estate of 100,000 acres, his warehouses and commercial interests. Cousin Dickson urged the Gourlays to give up tenant farming and go to Canada. Gourlay accompanied Dickson to Scotland and there bought from him another 500 acres adjacent to the 866 owned by his wife. In 1810 the challenge Deptford Farm presented was too interesting to relinquish, and he had barely begun his study of the English poor.[29] Dickson returned to Canada alone. In 1814 another of Mrs. Gourlay's cousins, the Honourable Thomas Clark, also a legislative councillor, had come with his wife to Deptford Farm. "Wounded and faint", they rested after three years of war in which the Americans had burned their house and wharves on the bank of the Niagara River.[30] For a second time Gourlay refused to leave, this time because his affairs were too unsettled. Cousin Clark helped him with a loan of £500, for he was intending to leave war-torn Canada and invest his money in Britain. Gourlay also declined an invitation from Morris Birkbeck whom he met in Bath. Birkbeck was inviting knowledgeable British farmers to buy land in a development scheme he was promoting in Illinois. He told Gourlay he would give him a farm with no down payment, but Gourlay declined.[31] He said he preferred to go to Canada where he had his own land, but the truth was that it was not his nature to follow the plans of another. For four years now, he had been studying W. Faden and D. W. Smyth's *Topographical Description of Canada* in his spare time and planning how to lay out land in the wilderness for a possible colonization scheme of his own.[32] Now that partial settlement of his lawsuit had been achieved, he

27Cobbett, *Parliamentary Debates*, XXI, Feb. 27, 1812.
28*Statistical Account II*, p. 494; *The Banished Briton*, #17, p. 180.
29*Ibid.*, #4, p. 41.
30*Ibid.*, p. 180; *Statistical Account II*, p. 504.
31*Statistical Account I*, ccxii ff.
32*The Banished Briton*, #17, p. 180.

had decided to make a quick trip to Canada, look over his land and return for his family if the prospect were pleasing.

He made a swift journey to Scotland for a reconciliation with his father. His father had purchased a copy of *The Village System* and had stopped listening to the derogatory reports that had caused him to disinherit his eldest son. He approved his son's aims for the British poor. Oliver could do nothing for him financially, as he too was faced with bankruptcy in this difficult postwar period. His affairs had been in the hands of administrators since December 16, 1815, when his estate was sold to the amount of £96,000. Thomas Henderson was also in financial trouble.

If only he could procure his money from the duke, Robert could leave for Canada immediately. He posted up to London where Lord Eldon again was intending to delay the order that would force the second peer of the realm of pay. Romilly, who knew the dilatory ways of the Court of Chancery, was pleading ineffectively against Eldon's suggestion to postpone the case for another six weeks. Rushing between Romilly and Eldon, Gourlay begged Eldon to realize that he was too hard up to wait any longer. Eldon was so surprised to hear a Georgian gentleman acknowledge that he was insolvent that he gave the "Long Order" that authorized Gourlay to collect his money. When Somerset still refused to pay, Gourlay "put an execution" on the duke's house and collected his money. The story of his interrupting the chancellor in the court was distorted as it went from mouth to mouth till it came back to Gourlay that he had threatened the Lord Chancellor of England violently. The authorities were told to watch Gourlay closely, for recently an attack had been made against the Prince Regent himself.[33]

On April 2, 1817, Gourlay was ready to leave for Canada. He was not the only reformer leaving England. On March 27 William Cobbett, to avoid imprisonment under the Gagging Laws, sailed for the United States after making secret preparations, "My choice," wrote Cobbett who would mail material for his paper from his self-imposed exile, "lies between silence and retreat."[34] The day before Gourlay crossed the road to Deptford Farm to take the coach to Liverpool, Mr. Fisher, who rented Allcannings Farm from John Benett, met Benett in the same inn to ask him to lower his rent or he would be ruined. Benett said he would evict Fisher and rack rent to future tenants. Benett was preparing to contest one of the Wiltshire county seats. He was defeated and left the polls under armed guard in one of the bitterest election contests ever known in Wiltshire, but the reformers gained no victory either.[35]

[33]*Appeal*, p. 76.
[34]*Salisbury Gazette*, April 10, 1817.
[35]*Kaleidoscopiana Wiltoniensia* describes the famous audit day at Deptford Inn, April 1, 1817; *Statistical Account I*, p. cccxxx.

Gourlay's position was better than Fisher's for he owned 1,300 acres in Canada. He had listened to the seeming ease with which his relatives, Robert Hamilton, William Dickson and Thomas Clark, had prospered in the new land. He was only thirty-nine years old, young enough to start afresh. A better life should lie 3,500 ocean and wilderness miles away in a country so new that surely there would be no "slippy stane" on its vast and empty doorstep.

10

O Canada!

When Robert Gourlay reached Liverpool with £92 in his pocket to take him to Upper Canada and back, he found that first class passage would cost him £36 15s 6d and he must supply his own bedding. His plan was to travel to New York, look over land in the states that banned slavery, take the quick American route up the Hudson River to Niagara and return by Quebec in the autumn. Embarkation preliminaries included finding some one to swear that he was not a mechanic, for mechanics, at this initial stage of British industrialization, were forbidden to leave the country. Before he could get his trunk on board, the ship of his choice sailed off in a favourable wind and the next one was bound for Quebec. He decided to enter Canada by Quebec and return by New York.[1]

While he waited, he received a letter from his wife with the news that her brother Thomas was threatening to sue them for £2,000 owing him on the note signed jointly by his father and himself. More than a little upset, he answered her: "If they push on thus, what will they get? Were I simply to be ruined by bankruptcy,—by losing all, it would be a light thought in comparison of remaining eternally under obligation. It dries up my life's blood, and makes me feel blasted forever." Then in a fit of testy masculinity, because he naïvely hoped to forget business affairs for a time, he added: "I wish you had kept this cruel news to yourself. Did I not say I had done all in my power; and would try to forget matters, which I could not mend. Do no more in my business; but it is unfair, my dear Jean, to be thus hurried with you. Do what you can without vexing yourself; and let us look to all as proper trials for us. As your brother has acted, and does mean to claim, it is, in fact, lucky that I have heard previous to sailing, that I may not deceive Mr. Clark.

---

[1] *The Banished Briton*, #17, p. 180.

How many people are there in the world perhaps even more hardly dealt with than us."[2]

He had time to visit a friend in nearby Chester. There he read James Melish's *Travels through the United States of America in the years, 1806, 1807, 1809, 1810, 1811,* a book designed to acquaint Irish immigrants with the United States. Immediately the idea struck him that if there were a similar book about Upper Canada, emigrants would choose to go there instead of the United States where they were lost to Britain. As his ship bellied its way across the Atlantic, he drew up a list of questions about Upper Canada. He patterned it after the questionnaire Sir John Sinclair had used for his recent *Statistical Account of Scotland,* modifying it with the knowledge gleaned from conversations with his wife's Canadian cousins, men like Morris Birkbeck, and from his study of Upper Canadian maps.[3] As he neared Canada, his mind was filled with optimism for there, he told himself, he would find a "land free from influence".

Only an optimist could approach even an underdeveloped country like Canada with such an expectation. No people were filled with firmer convictions than those who had founded the province where lay the land he was travelling to see. Upper Canada had been formed by an influx of 10,000 United Empire Loyalists who had left the revolting Thirteen Colonies for what they called loyalty to the British monarchy, for what the new republic of the United States termed inability to change with the times, and what posterity calls "the most reactionary element of the first empire".[4] This faction opposed the mélange of other opinions introduced by post-loyalist immigration and the inevitable pressure of changing events.

Upper Canada's government was determined by the Constitutional Act of 1791 that had been drawn up by Pitt's favourite cousin, William Grenville. More liberal Fox wished the new constitution to take into account the fact that Canada existed next to the country that had just revolted against Britain because it refused to have its growing economy hamstrung by imperial red tape. Tears came to Fox's eyes when he realized during the debate on the Constitutional Act in the House of Commons that he and his political friends had come to the parting of ways because of differences in ideology. The Constitutional Act, under which Canada would be governed for half a century, was Tory in composition.

---

[2]Liverpool, April 17, 1817, *ibid.,* #25, p. 304.

[3]In case this should seem presumptuous on Gourlay's part, it may be stated here that Edward Gibbon Wakefield had not been in Australia when he made his recommendations for scientific colonization. He based them on maps and theory.

[4]Chester Martin, *Empire to Commonwealth* (Clarendon, 1929), p. 61.

Under this constitution the governor-in-chief and lieutenant governors sought approval for major decisions from the faraway Crown and
colonial secretary in the old pattern of a royal colony. An answer to a
query could take a year to arrive, and legislation could be annulled as
long after as two years. The appointed executive and legislative councillors were torn between their on-the-spot knowledge of the local situation and their obligation to maintain the constitutional *status quo*. The
Constitutional Act had provided an assembly elected by fairly equal
ridings, but the relationship between the executive and the elected representatives was not defined. Executive councillor and Speaker of the legislative council Judge Powell was disinclined to ignore the assembly.
Executive councillor, rector of York, chaplain of the troops, and master
of the Home District school Dr. John Strachan deplored its existence.
The Honourable William Dummer Powell was an ex-American with
liberal leanings. Dr. John Strachan was a former Aberdonian of authoritative bent. By 1817 their differences of opinion were becoming marked,
and both were vying for the post of chief adviser to the lieutenant
governor in an incipient struggle for power in the capital at York.

Upper Canada's first parliament had been convened in September
1792, the month of the September massacres in France. Only an ill-
defined and undefended border separated Upper Canada from the new
United States of America. Fear that democracy and republicanism might
be forced on British North America by armed invasion from both France
and the United States was very real. Any one could be suspected of subversive activity. In 1795 one of the members of the Upper Canadian
assembly itself had been accused of sedition. This was loyalist Isaac
Swayze who had spied for the British in New Jersey, been captured and
sentenced to death. He had escaped to Canada by exchanging clothes
with his wife when she came to say farewell to him in jail.[5] He was
rewarded for his loyalty with choice land near St. David's in the Niagara
peninsula, but even this background was not enough to protect him.
At his trial for sedition Swayze was fined £10. Because of the lightness of
the penalty, it may be assumed that his criticism of the government was
the result of temporary disgruntlement with life in the new land, not
any change to radical political tenets.[6]

The best bulwark against encroaching ideas or troops was considered
to be a landed aristocracy. Lord lieutenants, one of whom was Gourlay's
relative, legislative councillor the Honourable Robert Hamilton, were
appointed to supervise the loyalty of the people and settle the empty
land that was an invitation to conquerors. On one memorable day, October 6, 1792, six townships averaging 60,000 to 70,000 acres each were

---

[5]C. Durand, *Reminiscences* (Toronto, 1897), p. 507.
[6]W. R. Riddell, *Upper Canada Sketches* (Toronto, 1922), p. 88.

promised to men who would bring in settlers quickly, and four more held for distribution.[7] The ordinary loyalist grant ranged in size from fifty acres for the head of a family and for each member to 5,000 acres for a field officer or chaplain. When the unsalaried executive and legislative councillors complained that military personnel were being favoured excessively, they were granted 6,000 acres. As they had easy access to the surveyors' reports, they naturally selected good land.

Acreage did not constitute prestige in early Upper Canada for land was as common as the blackberries on the bushes. The leading men were soon the merchants who achieved success in the fur and carrying trades. Possession of land was still desirable and they added to their original grants by taking land in payment for debts, buying it cheap from disbanded soldiers with no taste for farming, or from widows and orphans. Now that peace seemed assured, Gourlay's merchant relatives, Thomas Clark and William Dickson, were making large scale plans for quick development and sale of two townships. Government officials in York called them land speculators and equated them with Yankee land jobbers.

York officials, who lived on fees or salaries, envied what appeared to be the easy affluence of the merchants. They derived their importance from the administrative positions that were naturally centred in the capital. Newcomers applied to the land office for land grants. Settlers came to pay fees. The largest number of salaried positions was available in York, and government officials dispensed salaried appointments such as sheriff and postmaster. All were sources of cash in a country where money was scarce. The "place men" of Britain were notorious for their subservience to the ruling clique. The Upper Canadian government would soon be criticized for the same kind of control.

It was inevitable that the administration of the new country would incur criticism. Critics were dealt with in summary fashion. William Weekes had come from Ireland in 1798, the year in which young Gourlay and his cousin had looked over the rebel battlefield. In 1804 Weekes was elected to the assembly. Here he opposed the passage of the Seditious Alien Act on the grounds that it constituted a threat to the liberty of every person in the province.[8] He also demanded that the books of the Inspector of Public Accounts be examined in the assembly. An investigation disclosed no dishonesty, but it was clear that Lieutenant Governor Hunter was authorizing expenditures without consulting the assembly. The lieutenant governor considered this an expeditious way of clearing

---

[7]G. C. Paterson, *Land Settlement in Upper Canada* (16th Report of the Dept. of Public Records and Archives for the Province of Ontario, 1921), p. 42.

[8]Variously called the Sedition, Alien, Seditious Attempts or Seditious Alien Act, but herein the Seditious Alien Act as both facets are important to this biography. The British Treason and Sedition Act, 1795, was opposed by Fox and Sheridan, the American Alien and Sedition Acts, 1798, by Jefferson and Madison.

his desk. At an earlier period in history a British king had lost his head for ignoring parliament, but Weekes received no praise for bringing this matter to public attention in Upper Canada.

The following year Weekes gained an ally in the person of a newly appointed judge, Robert Thorpe. As Thorpe listened to his cases, he concluded that the average settler was often victimized. He formed the Upper Canadian Agricultural and Commercial Society to discuss current problems. Legislative councillor Richard Cartwright of Kingston accused the judge of demeaning his position by consorting with farmers and mechanics of low education and, *ipso facto,* of questionable character. He urged Lieutenant Governor Francis Gore to rid the province of men who "encouraged the Perverseness of ignorant people in the Matter of Politics".[9]

Weekes ceased his criticism rather suddenly. When the Honourable William Dickson objected to Weekes' assertion that Lieutenant Governor Hunter had allowed laxity in the land department, Weekes challenged Dickson to a duel. It was Weekes who was killed. Weekes' riding promptly elected Thorpe in his place. In an eloquent speech at the polls Thorpe pictured "the Shade of his departed Friend looking down from Heaven with Pleasure on their Exertions in the Cause of Liberty", and was elected, 268-159.[10]

Thorpe lost no opportunity of using his knowledge of parliamentary procedure and precedent to embarrass his superiors. Lieutenant Governor Gore moved him from the York circuit, which was filled with settlers of independent opinions, to the Quinte circuit, which was largely loyalist. There it was hoped that Thorpe would no longer find "some characters weak enough to be misled by the Arts and Cajoleries of that worthy disciple of Mrs. Wollstonecraft, his wife, which have now lost much of their effect in their present neighbourhood. . . . It was their practice at these convivial meetings to suggest first one thing and then another as a proper subject for Parliamentary Interference till they had hit upon something that their guests seemed to approve; who were by this means led to suppose, when these Subjects were afterwards agitated in the House of Assembly, that in approving measures grounded upon them they were supporting ideas of their own, when they were in Fact acting merely as Dupes of Mr. Thorpe in his Attempts to create confusion".[11] When the loyalism of the Quinte circuit was not sufficient to quench Thorpe and his wife, Gore engineered his recall to England.

9R. Cartwright Letter Book III, unabridged, Queen's University Library, p. 160.
10*Ibid.,* c. p. 165.
11*Ibid.,* April 2, 1807. Mary Wollstonecraft, wife of Henry Godwin, was author of *Vindication of the Rights of Women,* 1792. Thorpe successfully sued Gore for damages in England. Castlereagh to Gore, 19 June, 1807, Doughty and McArthur, *Documents relating to the Constitutional History of Canada,* 1791-1816, p. 310; *Canadian Archives Report,* VI, Note D, "Political State of Upper Canada in 1806".

The aversion of the government to free debate of public affairs was next challenged by the emergence of an independent press. Till now, the official *Upper Canada Gazette* had been the sole newspaper. Joseph Willcocks, who had also left Ireland after the collapse of the United Irishmen in 1798, founded the *Upper Canada Guardian* in 1807 for, in the course of his duties as sheriff of the Home District, he had formed a poor opinion of the administration. In 1808 editor Willcocks was elected to the assembly. There he moved that any member of the House should have free access to the minutes, for he wished to print some of the debates. Though the right of newspapers to publish parliamentary debates had been won in Britain in 1771 (John Wilkes and the famous No. 45 of the *North Briton* (April 23, 1763, were part of this struggle), the motion was defeated. Willcocks objected to the terms of the District School Bill for he said that the lieutenant governor would appoint no one but Anglicans as the eight district schoolmasters whereas dissenters formed the majority of the population. Provincial judges Sherwood and Boulton called the *Upper Canada Guardian* a pestilence in the land and Lieutenant Governor Gore reviled the editor as an "execrable monster who would deluge the province with blood".[12] They all combined to teach Willcocks a lesson. Willcocks was accused of acting in contempt of parliament because one evening in a private house he had criticized the actions of the first parliament of 1792. He was jailed for the remainder of the session. While he suffered the indignity of being confined in a cell "where every office of nature is obliged to be performed, there being no convenience", he wrote an Address to the electors of his riding: "It is the public, whose welfare and tranquillity are violently attacked, for what is my case to-day, may be the fate of every opposition member of any subsequent parliament; . . . a too ready submission, often occasions a repetition of the same evil".[13]

His *Address* sounded a good deal like Gourlay's pamphlet of the same year, and his treatment like that of Sir Francis Burdett, member for the borough with the electorate too large to bribe. Two years later Sir Francis would be sent to the Tower by the British House of Commons for championing Dr. John Gale Jones who had organized a debate outside the House of Commons on the disastrous Walcheren expedition. The Tower was a cut above York jail in the social scale of incarceration, but the object of both was the same—to silence a member of the opposition.

The man who moved in the Upper Canadian assembly on January 30, 1808, that "the expressions said to have been made use of by Mr. Joseph Willcocks are false, slanderous, and highly derogatory to the dignity of this house" was Isaac Swayze, the man who had been accused of

---

[12]J. C. Dent, *The Story of the Upper Canadian Rebellion*, I, p. 90; See also H. Guest, "Upper Canada's First Political Party", *Ontario History*, December, 1962, pp. 274-296.
[13]*Upper Canada Guardian*, Feb. 6, 1808, from *Statistical Account* II, pp. 655ff; Jls. Leg. Ass., OA Report #8, p. 225.

sedition in 1795 himself. Though Swayze now chose his words more carefully, his life was not lacking in colour. In January 1806, Swayze reported
that three men with blackened faces broke into his house and stole three
money bags containing over £500 in licence and land fees.[14] When no one
seemed to believe his story, Swayze withdrew from the assembly his petition to be excused from restoring the money. Four years later Willcocks
(*Upper Canada Guardian*, July 10, 1810), urged Gore to prosecute him
for rendering counterfeit bank notes.

Soon after this, Willcocks ceased publishing his paper, likely for lack
of money. It could not have been from lack of readers for Judge Powell
said a copy of the *Guardian* was in every home. Willcocks' opposition
tactics were continued in 1809 by a reforming pamphleteer, John Mills
Jackson, an immigrant from Wiltshire. In *A View of the Political Situation in Upper Canada*, Jackson attacked the leaders of Upper Canada
who, glutted with land themselves, raised the fees for a deed of grant
to nearly $40 for 200 acres and ordered the surveyor general to prefer
those who could pay fees fastest. He attacked the men who had caused
Judge Thorpe's recall. Thorpe was not a threat to the safety of the province. He had merely disclosed how the Mississauga Indians had been
cheated out of 17,000 acres by a false map[15], and how Dereham and
Norwich townships had been sold in the council room for one-third their
value.

In spite of all the attempted government censorship, the safety of the
province ended in 1812 when the United States declared war on Canada in
a last burst of bitterness against Britain. The war seemed to justify all the
measures the government had taken to prevent American ideas from
entering the province. Two members of the assembly, Benajah Mallory
and Abraham Markle, and former member Joseph Willcocks defected.
Friend suspected friend. One general threatened to burn the houses over
the heads of those militia men who did not answer his call, and another
seized sleighs with provisions that farmers were taking back to hungry
families. Tongues had to be guarded more carefully than ever. One member of the assembly, John Willson, said that the times were too dangerous
for a man to open his mouth. He went home, never to reappear for the
rest of the session.[16] Not all the terrors of war are bullets.

14Feb. 8, 1808, Upper Canada Sundries, PAC.

15This was actually a popular misconception. In 1805 under Lieutenant Governor Hunter
   (who died three weeks later) a provisional treaty was signed on the basis of an existing
   map that was later found to be inexact. His successor, Administrator Grant, awarded
   £700 additional compensation that did not satisfy all the Indians concerned.

16Jls. Leg. Ass., *Ninth Report*, OA, p. 340. News of the declaration of the war was
   despatched to Thomas Clark in Upper Canada by special messenger from John Jacob
   Astor in New York so that Clark could save their western trading interests. The
   messenger was punished for this slip in intelligence, but no one punished Clark and
   Astor.

In 1813 the Americans came billowing across Lake Ontario to attack York itself. The reverend Dr. John Strachan was the hero of the hour, for the main body of British regulars had just left to guard Kingston. It was he who demanded the signing of the capitulation terms that ensured the proper care of the prisoners and the cessation of looting. He himself rescued a friend's silver teapot from one of the jubilant victors, though it was understandable that he had not been able to save the provincial treasury of £2,144 and the parliamentary mace, nor prevent the burning of the parliament buildings.[17] In 1814 eight men were sentenced to death for treason by the twenty-three year-old acting attorney general, John Beverley Robinson, Dr. Strachan's foster son. Some rejoiced at justice triumphant, but others called it the Bloody Assize of Ancaster.[18]

The war ended with a peace that was a stalemate in every respect but one—Canada would remain British. Even so, Britons contemplating emigration did not automatically select Canada as their destination. They were attracted to the fast-moving United States. Long before he decided to see his land in Canada, Gourlay had deplored the numbers being lost to Britain. Now, as he crossed the ocean, he was regarding his project for a statistical account of Upper Canada with his customary enthusiasm. Cobbett had written in 1815 that he was disillusioned with his seemingly ineffective efforts to change human nature and achieve reform. Gourlay was not discouraged. He was certain that he could give many of England's poor a new start in the new land.

He was about to enter a growing province eager to get ahead now that the war was over. Recently departed Lieutenant Governor Gore had reported to Lord Bathurst, secretary for war and the colonies, that he had left Upper Canada in a "spirit of union and conciliation".[19] Gore had prevaricated. Gore had been sent out to govern in the eighteenth century sense of the word, but many of the men who comprised the elected assembly provided by the constitution of 1791 were tired of being ciphers. A confrontation between elected and appointed bodies was inevitable. On April 3, 1817, the day after Robert Gourlay left Deptford Farm, Francis Gore had prorogued the Upper Canadian assembly that he considered, in official language, was exceeding its prerogatives, or in the vernacular, was getting too big for its britches. Executive councillor and Speaker of the legislative council Judge Powell had advised him not to prorogue the assembly, but a lieutenant governor was under no compulsion to listen to colonial advisers, either appointed or elected.

---

[17]J. C. Dent, *The Canadian Portrait Gallery*, I, p. 105. The main account of the battle of York is Strachan's version. See C. W. Humphries, "The Capture of York", *Ontario History*, 1959.

[18]W. R. Riddell, "The Ancaster Bloody Assize of 1814", OH, 1923. The sentence was hanging till half dead, then taken down and drawn (disembowelled) and quartered while still alive.

[19]Gore to Sherbrooke, York, May 20, 1817, Upper Canada Sundries, PAC.

Despite Gore's rosy report to Lord Bathurst, Upper Canada was not in a docile mood. It had passed out of its initial stage of concentrating on mere survival. It was a quarter of a century old. It had survived a three-year war that had tempered in it a pride of being. It had also retarded natural progress. Business men from one end of the province to the other were writing each other, impatient about current postwar stagnation. Some resulted from policies dictated by a faraway government ignorant of local affairs and divided itself on the advisibility of maintaining a luxury like an empire. Some was caused by local jealousy. Till the arrival of a new lieutenant governor the head of the government was Administrator Colonel Samuel Smith who was prepared to be a caretaker, not an originator. Three forces were ready to pull at him—liberal Judge Powell, tory Dr. John Strachan, and two prominent Niagara business men, legislative councillors Thomas Clark and William Dickson. Upper Canada was no longer the uncomplicated colony that Gore had described to the home government and that Gourlay was expecting in a country so young.

II

Coming to the Boil

Robert Gourlay disembarked at Quebec on May 31, 1817. The size of the
St. Lawrence River had a breath-taking effect on the man accustomed
to the little rivers of Britain. "Take up the map of the world," he rhap-
sodized as he went along, "and you will not find upon its surface any
thing to compare . . . in point of grandeur, to the waters of the St.
Lawrence." This river could carry the poor of England to a better life.
As they were, the poor were only a burden to themselves and to England.
As consumers of British exports and growers of food for Britain, they
would bring prosperity to both countries. With his usual optimism, Gourlay saw no reason why the government could not plan for and settle
50,000 people a year in these great empty spaces.[1]

When he set out to explore Quebec, he was immediately struck by
its miserable planning. It was worse than the Old Town in Edinburgh
for some of its streets were only fifteen feet wide. He made mental notes
for replanning Quebec along the same clean lines as the new section of
Edinburgh.[2] He called on John Neilson, a liberal publisher and bookseller to discuss his proposed book of statistics, but Neilson was not at
home. On the third day he took the steamship *Accommodation* to Montreal for he had no time to linger if he were to visit Niagara, Detroit, Ohio,
Philadelphia, New York, and return to Deptford Farm in October.

On the boat he had missed in Liverpool Gourlay had sent a letter by
the quick Hudson River route to his wife's cousin Thomas Clark to meet
him in Montreal but when he arrived, there was no message for him. As
he waited, he made inquiries about a new settlement in Upper Canada
at Perth, the first attempt of the British government at planned immigration since the influx of the loyalists. When the British government was
faced with the problem of disbanded soldiers after the Napoleonic war,

---

[1]*Statistical Account I*, p. cccxiii.
[2]"Plans for the Improvement of Boston", *The Banished Briton*, #8, p. 80. Gourlay was
also keeping a "daily book".

it decided that a cheap solution was to give them free land in Upper Canada. A government advertisement promised free passage, 100 acres of land, food for six months and free implements. Four townships were laid out for them near Perth. When Clark did not arrive after he had delayed eleven days, Gourlay decided to use a letter of introduction he had asked Clark to give him to the head of the settlement, Captain Fowler.

He wrote a long letter to his wife, and then set out on foot for Perth. A walking tour ought to complete the restoration of his health after the nervous strain of his endeavours in Wiltshire, and travelling on foot was the best way of ascertaining how the people lived. Along the rutted road beside the great river he strode with that animated half gallop that Scotsmen call a walk, flinging out his long legs and flourishing the crooked stick that had helped him up many a Scottish hill. It was the middle of June. The bone-warming Canadian sun roistered in the sky all day, and then set in a blaze of crimson glory.

As he walked along, he surveyed the canal needs of the St. Lawrence, for the success of any immigration scheme would depend on transportation. On this gigantic river there were only two locks. Two rival merchants who had spent about £300 at the Moulinette rapids had petitions before the Upper Canadian assembly at the moment awaiting a solution of their claims.[3] Gourlay had studied the plans for Scotland's Caledonian canal. Now he marvelled over the slow development of canals along the great river that was Upper Canada's chief hope for progress. The problem was partly lack of planning and partly lack of manpower, he reasoned, and he could remedy both conditions if he had the chance. Twenty pounds would bring over a family of three. The first shipload could begin work the week of their arrival, and they would smooth the way for the next in an endless chain of prosperity.

He could see it all so easily. This country was bulging with riches. Like a squire estimating the yield of his acres, he calculated the value of Upper Canada to be about six million pounds. Britain had floating capital to invest. Who with property worth so much would hesitate to borrow money to develop it? The valued rent of Scotland alone was six million pounds of which his own shire of Fife produced £378,755. "Should I stickle at borrowing two millions to improve its value three times over in five years, which most certainly could be done with proper management" said this man who was bold enough to undertake it.[4] The little torment of Deptford Farm seemed puny and far away.

By now Gourlay had passed the French seigneuries with their farms laid out in strips fronting on the river and was in Upper Canada. In

3Feb. 24, 1817, Jls. Leg. Ass., OA, Ninth Report, p. 324.
4*Statistical Account I*, p. ccccxiv.

twenty years there had burst into Upper Canada about 92,000 people, a population equal to that of Lower Canada after 200 years of existence. Soon he reached Cornwall, still called Johnstown after loyalist Sir John Johnson who settled his King's Royal Regiment there in 1784. The first grammar school in Upper Canada was opened in Cornwall. Merchant Richard Cartwright, former partner of Robert Hamilton, had written to Dr. George Hamilton, Jean Gourlay's uncle in Gladsmuir, to recommend a tutor for his sons. A young schoolmaster-minister named John Strachan accepted the position and emigrated to Canada in 1799. He was curious to see foreign parts. His chances of obtaining a living in Scotland were slim, and the salary offered was £80. He had spirit and ambition. After three years of being a tutor, he opened a school. The year after, he became Anglican curate in Cornwall with a salary remitted from England. In 1807 he married Ann Wood McGill who received a £300 annuity from the estate of her late husband, Nor' Wester Andrew McGill.[5] Two years later he wrote to Aberdeen requesting an honorary degree in absentia. "An Honorary degree . . . would be of gt. service to me here, for altho' there are no distinctions of rank in this country, no people are so fond of them. . . . I have no doubt, but that a degree might in some measure increase my influence".[6] The doctorate in divinity was sent across the ocean in 1811.

Just before the War of 1812, Strachan left Cornwall to become headmaster of the Home District Grammar School and chaplain of the troops in Fort York. He was Anglican, and royalist. He came from Aberdeen which had remained a pocket of episcopalianism and royalism in Scotland. He wrote a pamphlet praising the king in such extravagant terms that even his foster son and pupil, John Beverley Robinson, laughed at the whitewashing as he described it to his friend John Macauley in Kingston.[7] He was anti-American. Americans were vain, rapacious, and without honour, he said. He disapproved the necessity of sending Canadian youth to school in the States where they were taught "anarchy in Politics and infidelity in religion". He equated republicanism with democracy with atheism. He tolerated an elected assembly because he must, but as early as 1808 he was scheming how to control it: "By and bye my pupils will

---

[5]This may have contributed to Strachan's opposition to Lord Selkirk's settlement at Red River. Strachan to Dr. Brown, York, Dec. 1, 1818, G. Spragge, ed., *The John Strachan Letter Book*, p. 183. Lord Selkirk was the brother of Whig Lord Daer who had been brave enough to hold a birthday celebration for Fox in Edinburgh a decade before.

[6]Strachan to Dr. Brown, Cornwall, Oct. 8, 1808, *ibid*, p. vi. Strachan's academic qualifications were an M.A. in 1797, University of Aberdeen. A B.A. was customarily awarded at seventeen, and an M.A. at nineteen. Strachan studied part time for divinity during the two years he taught at Denino. Four years full time and six years part time were the requirements for a divinity degree at St. Andrews.

[7]Robinson to Macauley, York, Nov. 25, 1810, Macauley Papers, OA.

be getting forward, some of them perhaps into the House & then I shall have more in my power".[8] John Strachan's aims for Upper Canada could not be anything but pure, for he would raise the moral tone of the province according to Anglican principles and preserve it loyal and unchanging to the British crown.

After a brief pause in Cornwall, Gourlay walked on to Brockville. Here he turned away from the St. Lawrence to head forty miles inland along the road that had just been hacked in to Perth. He intended to begin here his inquiries for his proposed statistical account of Upper Canada, and formulate immigration plans of his own that would avoid any mistakes made at Perth. He took a questionnaire to each homestead in turn. About fifty heads of families signed statements for his chart. They were disgusted with the amount of money they had had to spend waiting for a boatload to gather after they reached Greenock. Many returned home, unable to finance themselves. The 700 who persisted arrived in Upper Canada while civil and military authorities were still quarrelling over the placement of "Mr. Goulburn's vagabonds" and had to be fed on government rations during the winter. Otherwise they were satisfied with their progress.[9]

Gourlay was pleased with Perth's strategic location near transportation on the St. Lawrence. He asked Captain Fowler if he could be granted a large tract nearby, for he could see the advantage to be gained from proximity to an existing settlement. Fowler told him that the limit in the Perth area was 100 acres, but he could buy up separate lots and wait for the land to increase in value. As Gourlay's object was development, not speculation, he could see that this was not the place to begin his endeavours.

On July 2 he reached Kingston where he presented a letter of introduction to John Kirby, merchant and member of the assembly. Kingston was a busy little place with its population of 2,250 plus 1,500 soldiers, its seventy-eight taverns, its churches, Dorcas Society, Female Benevolent Society, its courthouse, jail, market place, its newspaper (the *Kingston Gazette*), its Anglicans who "sneared" openly at the Presbyterians, and "Cream Street" where the cream of its society was building large new homes. Here merchants like Richard Cartwright profited from military contracts and the spending habits of the soldiers.

Kingston had been disappointed when York had been selected as the site for the provincial capital. Possibly the town should have been

---

[8]Strachan to Dr. Brown, Cornwall, Oct. 9, 1808, Spragge, *The Strachan Letter Book*, pp. 59, viii.
[9]*Statistical Account I*, pp. cccx-cccxii; II, pp. 561, 662; Playter, "An Account of the Founding of Three Military Settlements", with notes by E. A. Cruikshank, OH, XX, p. 98.

happy, for by this rejection it escaped some of the calumny that was heaped on the capital whenever the government failed to satisfy the expectations of a diverse population. There were those in the Kingston area who had specific grievances that could be remedied only by the central administration at York. One of these was yeoman Amos Ansley of Lot 12, first concession of the township of Kingston, who had come as a loyalist with Captain Michael Grass when the land was "a howling wilderness". Amos had been complaining fruitlessly about the boundaries of his farm till finally in 1800 he wrote to surveyor general D. W. Smith in York: "the deeds now cut our farmes to peases and make a great confution . . . and if the Lines are to Be altered from the original Survey to agree with the deeds I must loose a valooable farme and many of my Neibours are in the same situation".[10] Amos was not the only complainer, but as he was the most persistent, York officials scorned him as a chronic malcontent.

After he had explored Kingston briefly, Gourlay left by steamship for the Niagara peninsula. An overnight trip brought him to the warm greeting of the Honourable Thomas Clark. Clark's forty-room stone mansion overlooked the Niagara River just before it thundered over the Falls. The magnificent trees of the natural forest landscaped his grounds. Below the house at the foot of the steep bank eroded by the "Thunderer of Waters" was his busy wharf where furs from 1,500 miles inland were loaded for their trip along the Portage Road. Just below Niagara Falls at Queenston Gourlay at last met his wife's Hamilton cousins. The great house from which the late Honourable Robert Hamilton had dispensed princely hospitality had burned down in the late war when it was hit by hot shot. His son had built a new house facing the ruins on the promontory overlooking the rapids below the Falls. Then in Niagara (now Niagara-on-the-Lake) Gourlay renewed his acquaintance with the Honourable William Dickson, the man who had killed William Weekes in a duel, and who had been taken prisoner when the Americans burned and looted their way across the peninsula in 1813.

All three families of Mrs. Gourlay's relatives had suffered great losses in the war, for they had much to lose, but their powers of recuperation were considerable. The three families were among the most prominent in the province. This was just the company that Robert Gourlay loved, men who governed, men of consequence. He settled happily into Clark Hill where he proposed to enjoy a two or three-week visit before setting out to see his land in Dereham Township.

He had planned his journey with no consideration for the minute but

---

[10]W. Riddell, "Humours of the Times of Robert Gourlay", *Transactions of the Royal Society of Canada*, 1920, vol. 14, p. 70.

mighty North American mosquito that bred in the low-lying ground around Niagara. The brash newcomer was bitten so severely that he was ill in Thomas Clark's house for six weeks.[11] As he convalesced, he listened to Thomas Clark, William Dickson, and Robert (Junior) and George Hamilton carry on endless discussions of Upper Canadian affairs.

Foremost in Dickson's mind was the slow sale of 94,035 acres of the choice land he owned ninety miles west of Niagara at the junction of the Grand and Nith rivers. He had been negotiating the purchase with Clark when war interrupted their plans. As soon as possible after peace was made, Dickson completed the transaction, for he was eager to add land developing to his political and legal interests. On July 3, 1816, he paid Clark £15,000 and a few days later set out for the forks of the Grand. In the summer of 1817 the saws of a new mill had already screeched their way through lumber for the forty families he had settled expeditiously and providently. To date he had spent £4,000. He needed more settlers. He would take them from Scotland where he was advertising in the papers, or he would take them from Upper Canada's recent enemy, the United States. Americans were desirable settlers who could swing an axe the day they arrived. They were eager to purchase, for New England was already crowded and Upper Canada was closer than Ohio or Illinois.[12]

Clark too was ready for action. Even after Dickson's enormous purchase he possessed 29,000 acres in Nichol township, part of Block Four of former Indian land. Richard Beasley (Richard Cartwright's cousin) was anxious to develop what remained of his £8,887 purchase of 94,012 acres of Block Two after his £10,000 sale of 60,000 acres to Mennonites in Waterloo Township. Robert Nichol, a Scottish cousin of Clark's, quartermaster general during the recent war, partner with Clark in Nichol Township, member of the legislative assembly, also wished to develop immediately. Current imperial policy was preventing their plans for vigorous expansion.

On January 10, 1815, Lord Bathurst, secretary for war and the colonies (1812-1827), issued a directive to governor-in-chief General Sir George Drummond restricting the entry of Americans for fear they would alienate Canada from the Mother Country. When Lieutenant Governor Francis Gore returned to Upper Canada in September 1815, after the termination of the military interregnum, he reported that in spite of the directive Americans were pouring into the province.[13] Some were returning

11 *The Banished Briton*, p. 238. It is not clear whether Gourlay was allergic to the bites or if they aggravated his nervous condition.
12 J. E. Kerr, "Sketch of the Life of Hon. William Dickson", Niagara Historical Documents, #30, 1917; C. M. Johnston, *The Valley of the Six Nations* (Champlain Society, 1964).
13 Bathurst to Drummond, London, Jan. 10, 1815, PAC, Series G; Gore to Bathurst, York, Oct. 17, 1815, CO 42/356/123.

because they regretted their precipitate departure during the war, like John Thatcher, a volunteer who resented being cursed by an officer after the Detroit campaign. Others were simply looking for jobs or land. Gore was certain their purpose was to form a fifth column.

The only statute under which Gore could order the Americans out of the country was the Seditious Alien Act of 1804, but that would necessitate tedious prosecution. A better way was to prevent them from acquiring the land that was their chief reason for entering. On October 14, 1815, with the approval of the executive council, Gore addressed a circular to the commissioners authorized to administer the oath of allegiance (required for owning a land or business). The circular enjoined them to report to the office of the lieutenant governor all new entrants to the province. They were not to administer the oath "to any Person not holding Office in the Province, or being the son of a U.E. Loyalist, without special authority, in each case, from this office".[14] This angered the land-rich Niagara and Hamilton business men. Commissioner Dickson claimed that Gore's action was unconstitutional for it had not been approved by the legislature. He continued to administer the oath to incoming Americans till Gore ended his defiance by removing him from his post as commissioner on April 8, 1817, the day after he prorogued the assembly. To save face, Dickson thanked Gore for omitting his name from the commissioners' list because of the weight of his duties as legislative councillor.[15]

Dickson's complaints to Gourlay as he lay convalescing belied the contents of his letter to Gore. If the order to exclude Americans had been accompanied by intelligent aid to British settlers, the land developers would have had no cause for discontent. Instead, when the British government counted the cost of the Perth settlement, it announced that it would give no more aid to civilians, who must emigrate blindly. By the summer of 1817 immigration to Upper Canada practically ceased. Americans were excluded, Britons unencouraged, and legitimate land developers were called speculating thieves. The country could not prosper without an increase in population to share the burden of development.

The sole speculators were not the Niagara and Hamilton developers who were prepared to give good value for the price they asked for their land—mills, roads and a market provided by people living in proximity. Their equal in speculation was the government that had granted vast

---

14Minutes of the executive council, Oct. 7, 1815; *Statistical Account II*, pp. 416, 427.

15Dickson was partly correct in saying that the lieutenant governor could not prevent people from taking the oath of allegiance and oath of intention to reside in the province, and partly wrong for they had not been in residence for the seven years required for holding lands as specified in 13th George III. Bathurst to Smith, London, Nov. 30, 1817, PAC G58, p. 156; Dickson to Cameron, Niagara, April 27, 1817, PAC, Upper Canada Sundries; *Statistical Account II*, pp. 439, 522.

quantities of land to people who were loyal, military or official. Many of them were unable to develop it or were waiting for it to increase in value through the efforts of the average settler who broke his back and buried his wives as he cleared it. The government itself held idle two-sevenths of each township as crown and clergy reserves. The crown reserves were designed to render the executive government independent of the elected assembly and the clergy reserves were intended for the support of a Protestant clergy instead of tithes. Settlers were required to perform settlement and road duties, but the lands of absentee owners and the crown and clergy reserves lay roadless, idle and obstructive.

The original intention was that the four million acres of crown and clergy reserves were to be sold or leased to provide revenue, but the present government policy rendered unsaleable not only the land of the so-called speculators but also its own and the church's as well. The church could rely sometimes on charity to fill its coffers. On April 28, 1817, this advertisement appeared in the *Salisbury Journal*: "The Church in Canada Needs Money. The Lord Bishop of Quebec, Dr. Strachan of York, Dr. Stewart of St. Armand, will accept donations. £1100 already collected". The crown, in the person of the lieutenant governor and his advisors, was finding it more difficult, for as the colony increased in population, more money was needed for administration. This would have to be voted by the assembly, but Gore had just prorogued it with its business unfinished.

The trouble had begun in the last days (May 1816) of the parliament elected in 1812 before the declaration of war. On the surface, all was amiability. The assembly voted £3,000 to buy Lieutenant Governor Gore a service of silver plate and provided him with an aide-de-camp.[16] They voted a salary for a provincial agent to reside in London to represent Upper Canadian interests. They voted £500 to retiring Chief Justice and Speaker of the legislative council Thomas Scott, and payment to Judge Powell, his successor, for 2,183 land claims he had reported on, for they were outside his duties as judge. They voted £21,000 for roads and £6,000 to institute the system of elementary schools under the new Common School Act of 1816.

Land fees, licence money and even the £10,846 Lower Canada owed for duties collected at Montreal and Quebec during the war would not cover such extensive spending. For the first time in the history of the province, the lieutenant governor asked the assembly to vote money for the growing civil list. Salaries of higher officials had been paid since 1791 by a yearly grant from Britain of approximately £7,000 and any deficit appropriated from the "military extraordinaries". The military chest was

---

16*Jls. Leg. Ass.*, Mar. 29, 1816, OA, *Ninth Report*, pp. 297-9.

empty because of pensions to war disabled, widowed and orphaned. The assembly voted £2,500 to cover the civil list. Robert Nichol, member for Norfolk and chairman of the Select Committee on the Public Accounts, had pointed out several years before that money intended for those accounts still remained in the treasury. He pointed it out again, but the session closed, still amiable.

The ears of many assembly members, however, were blistered when they presented themselves for re-election to the seventh parliament. The £3,000 "Spoon Bill" was the main target. The extravagance of British officials in contrast with economy-minded local men was one of the reasons for the American Revolution. The sum of £3,000 was small in comparison with the £200,000, etc., etc. that Wellington was given after Waterloo, but it was astronomical when compared with the total Upper Canadian budget of £60,000. Few begrudged the money for a provincial agent in London, but when Gore appointed his own secretary every one knew who would control the agent. In addition, would the next money that the new parliament might be asked to vote to cover public accounts be dispersed or join the amount that Robert Nichol had shown was being held in reserve?

The seventh parliament convened on February 4, 1817, after three years of postwar depression had steeled some of its members to demand action. The situation in Upper Canada was aggravated by the same weather that had rotted the potatoes in Britain. In the cold wet summer of 1816 men had worked in the fields of Upper Canada with their greatcoats on. Five hundred and fifty families in the Glengarry district were petitioning for aid. Militia men who had been promised 200 acres of land for war services had not been given them. Claims for war losses had not been settled. The Niagara District had lost heavily but York had sustained only one per cent of the losses.[17] Bitterness was growing about the effects of the exclusion of American immigrants, the idle crown and clergy reserves, and the fact that absentee owners of wild land paid no taxes.

James Durand, who had been elected to deserter Markle's seat, voiced some of the discontent in an "Address to the Independent Electors of the County of Wentworth", February 14, 1817, in the St. David's *Spectator*. He objected to the delay in the payment of prize money promised to the young men who had captured Detroit in 1812. He claimed equal credit for the Common School Act of 1816 that had come to rest on "the Atlantean shoulders of Dr. Strachan of York". He reiterated his opposition to an amendment to the District School Bill by which £1,000 yearly was to be set aside to support a few students in divinity whose teachers,

---

17C. W. Humphreys, "The Capture of York", OH, 1959, p. 15.

noably Dr. Strachan, were to be paid whether or not there were students. He emphasized that he had opposed the suspension of habeas corpus and the imposition of martial law during the war, for it "closed the lips of most of our people . . ." and compelled "the people to bear the kicks and cuffs of those who are used to overbearing tyranny". He claimed that the House was increasingly prodigal with its large funds. He ended his Address with a fervent appeal: "Once more, then, My Friends, I invite you to favour me with the Honour of your suffrages; and by a long, strong, bold pull at one time convince the tools of corruption that the path to the people's patronage is honest, independent conduct".[18]

Durand's outspoken words horrified the government. Robert Nichol, given to criticism himself, moved in the assembly that Durand's Address was a "false, scandalous and malicious libel", that it reflected on the conduct of the lieutenant governor and the late parliament. The printer of the *Spectator*, Richard Cockrell, was ordered to appear before the House. The assembly considered asking Durand to apologize publicly in the *Upper Canada Gazette* or declaring him ineligible to sit in the assembly as his name did not appear on the assessment rolls. The latter punishment was unconstitutional, for Grenville's Constitutional Act had not specified that this was a condition of candidacy. It was finally voted to commit him to the common gaol in York till the end of the session for contempt of parliament. Sir Francis Burdett had submitted to the Tower in 1810 and Willcocks to York gaol in 1808, but Durand refused to sample the hospitality of York gaol. While the assembly was dealing with the establishment of a Bank of Upper Canada with a capital of £100,000, Durand skipped back home. The assembly voted by a close majority of one to expel him for the remainder of the session.[19]

Dr. Strachan's supporters in the assembly re-introduced the District School Bill amendments to provide funds for divinity students in a seminary of which Strachan would undoubtedly be principal. The legislative council rejected it; only two of its members were Anglican. The assembly introduced an Absentee Bill to tax lands owned by absentees and make them perform road duty. The legislative council rejected it; they owned more land than the members of the assembly.[20] Three commissioners were appointed to negotiate Upper Canada's share of the duties collected at Côteau du Lac.

The assembly voted £8,281 to cover the 1817 civil list. Two weeks later a committee asked the lieutenant governor about the disposition

18*Jls. Leg. Ass*, OA, *Ninth Report*, pp. 336-343.
19*Ibid.*, pp. 348, 353, 357.
20*Ibid.*, pp. 384, 389.

of the £2,500 they had voted for the same purpose the previous year.[21] Ten days later Gore replied that he had the Prince Regent's command to pay retired Chief Justice Scott a pension of £800 sterling and the remainder of the money "was subject to the disposition of His Majesty". Translated from officialese, this meant that what had been done with the rest of the £2,500 was none of their business. The power of the purse—the main peaceful weapon a parliament possesses against autocracy—did not lie in the hands of the elected representatives of the people.

The answer to the question to Gore was known before it was asked for Robert Nichol was still chairman of the Select Committee on Public Accounts. Therefore it was no accident that Nichol was ready with a list of eleven censorious resolutions to present to the assembly as soon as it had received Gore's written refusal to allow them jurisdiction over expenditures. Immediately (April 3) the House went into committee of the whole to consider the state of the province. Nichol's resolutions dealt with every matter of current importance: the exclusion of Americans from the province, the efficiency of the post office, the crown and clergy reserves, the 200-acre grant to militia members and settling their arrears of pay, estimated at £28,784, and immediate payment of war losses, estimated at £229,650.[22] If these resolutions were adopted, Nichol claimed that the Mother Country would be relieved of the burden of providing for the civil establishment, and the province would be boosted towards prosperity.

Quickly it was resolved that William Allan, postmaster at York, was to report in two days to the House, and that an address was to be presented to the lieutenant governor requesting humbly if any orders had been received from Britain for granting land to the Volunteers and militia. The first of four resolutions on immigration was adopted and the assembly adjourned till April 5.

On April 5 the address to the lieutenant governor regarding militia land received its three readings. The second and third resolutions on immigration were adopted, 13-7. The fourth was brought forward: "That the said Acts were still in force, and that subjects of the United States may lawfully come into and settle in this Province, hold lands, and be entitled to all the privileges and immunities of natural born subjects therein, on complying with the several formalities required by the said Acts and existing laws of this Province". The House adjourned till Mon-

---

[21]*Ibid.*, p. 358, 380. As the motion was moved by Jones and seconded by Vankoughnett, it is plain that conservatives as well as reformers were insisting on the right of the assembly to control money bills.

[22]Losses in both Lower Canada and the United States had already been compensated. E. A. Cruikshank, "Postwar Discontent in Niagara", OH, 1933, pp. 23, 24.

day. The speed with which the first resolutions were adopted augured well for the rest of Nichol's presentation.

Before the minutes could be read on Monday, the assembly was summoned to the Bar of the legislative council. There Major Francis Gore gave his assent to the bills he favoured, reserved the Bank Bill for His Majesty's pleasure,[23] thanked them for the bill of supply voted to cover the deficiency of funds, and informed them that because their "longer absence from your respective avocations must be too great a sacrifice for the objects which may remain to occupy your attention . . . I have, therefore, come to close the Session, and permit you to return to your homes. . . ."

The assembly could read between the lines. It had been Gore who had rid the country of individual critics of his administration, the Wyatt-Weekes-Thorpe faction, and now he rid himself of a whole legislature by proroguing it. His action roused such surprised indignation that certain members thought of impeaching him.

For his part, Gore considered that Colonel Robert Nichol had acted as he did, not through concern for the public good, but because of personal pique over failure to obtain the £27,000 he claimed was owing him for war and business losses.[24] Gore, and at least one of his advisers, executive councillor John Strachan, considered that Nichol had been instigated to introduce the resolutions by William Dickson and Thomas Clark. It was part of the constant pressure game of business rebelling against government restrictions, and government claiming that business practices militated against the larger aspects of government policy, in this case the security of the country against Americans.

As soon as he had announced the prorogation, Gore wrote to Lord Bathurst at the Colonial Office justifying his action. He had prorogued the legislature rather "than wait until such dangerous Resolutions, reported and adopted, should be promulgated to the Public, through the medium of the press." Nevertheless he urged Bathurst to authorize the immediate payment of war losses out of army funds as there was no money in the civil treasury. He did not regret his opposition to the admission of Americans, and he knew where the demand originated. "The interruption of the flowing immigration from the United States of America was particularly offensive to certain Land Speculators, who had become possessed of vast Tracts of Land under the injudicious sales made by the late President Russell. . . . If the restraining of Emigrants from the United States, from the settling of this Province, should be abandoned,

23OA, Report #9, p. 424; S. F. Wise, "Kingston Elections and Upper Canada Politics", OH, December 1965, pp. 209 ff. The Act's time limit expired before the king replied, thereby depriving Kingston of a chartered bank.
24E. A. Cruikshank, "The Public Life and Services of Robert Nichol", OH, 1922.

the next declaration of Hostilities by America, will be received by Acclamation, and the Loyal Population of the Colony will be reduced to defend themselves against the disloyal. . . .

"I restrain myself at present to the Assurance, that if early attention is not paid to compose the spirit arising, by the machinations of Land Speculators in this Province, the King's Government will be exposed in all future time, to purchase tranquility by the disagreeable measure of stifling sedition by rewards, and thus encouraging the evil."[25]

The next day he presented his letter to Bathurst to the executive council for approval. Dr. Strachan approved, for he regarded Nichol's resolutions in the assembly as "pregnant with revolutionary spirit".[26] Legislative councillor Dickson was asked to appear before the executive council to justify his continuing to administer the oath of allegiance to Americans. It was at this point that the council voted to remove Dickson's name from the list of commissioners. A fourth dimension was being added to the existing triangle of governing bodies—local jealousy. York and Niagara groups were now in opposition.

Upper Canada, with its widespread population and it growing economic needs was beginning to be too complicated for a major to administer. In the middle of May Gore departed for England with great relief, leaving events coming to the boil.

As Robert Gourlay lay convalescing in Clark's house during the summer of 1817, he heard Clark, Dickson, and Robert Hamilton Junior, expressing their opinions of the present administration. As his strength returned, Gourlay announced his intention of going to York to receive official endorsation of his project for a statistical account of the province, and to petition for land for the rather large immigration scheme he intended to propose to relieve the poor of Britain. Dickson said he would waste his time by going to York for the officials there would do nothing. He would get quicker results if he went straight to the colonial office in London.[27] Two Upper Canadians were already planning that very move— assembly member Colonel Robert Nichol to obtain compensation for his war losses, and Colonel Thomas Talbot in order to retain his control over five townships he was developing in the western section of the province. The inconclusive state of mind that ill health engenders still lay upon Gourlay, but insofar as he was capable of deciding, he had determined not to take Dickson's advice but to go to York as soon as his recovery was complete.

---

[25]Gore to Bathurst, York, April 7, 1817, Q322, p. 129.
[26]Strachan to Mountain, Lord Bishop of Quebec, York, May 12, 1817, G. Spragge, ed., *The John Strachan Letter Book*, p. 150.
[27]*The Banished Briton*, #19, p. 209.

12

To the People of Upper Canada

For the first time in his life Robert Gourlay was dispirited. He had left the old country beset by economic depression only to find the same in the new. His unexpected illness had forced him to accept Thomas Clark's hospitality long past the two weeks of his original intention. He could not steel himself to ask Clark for a second loan. Though it seemed that he had borrowed light-heartedly enough from his father in 1809, from Clark in 1814 and from Wilson in 1816, he had done so knowing that he had a twenty-one year lease, that Clark intended to invest his money in Britain, and that Wilson's money was a windfall. The role of suppliant did not sit lightly on his shoulders now. Delicacy prevented him from requesting a loan while he was Clark's guest. He set off for Niagara Falls and wrote Clark from there on September 1, 1817: ". . . I have every day had the strongest desire to converse the matters over with you; . . . A certain flutter of the spirits continually obtrudes itself with every desire to bring to an issue a question which must determine the course of my future life, and the fate of my family."[1] He listed his assets and liabilities and asked Clark to regard a loan on a strictly business basis. Clark manifested every inclination to assist him, but he had no ready cash till he received compensation for war losses or sold some land.

Gourlay then wrote to Governor-in-Chief Sir John Sherbrooke in Quebec to see if he could be granted more than 100 acres near the Perth settlement, for his own land in Dereham Township would cost more than it to develop. The answer was that regulations specified a uniform grant of 100 acres. While he awaited a reply, Gourlay set up his findings at Perth in a "Statistical Table showing the Commencement and Progress of Improvement, in 13 months, of the Emigrant Settlement at Perth". He sent copies to his wife for publication in British papers. How the gentle-

---

[1]*The Banished Briton*, #25, p. 305.
[2]*Ibid.*, #34, p. 466.

men of the Bath Society regarded Gourlay's report in the *Salisbury Journal* of November 24, 1817, is not known.

By now his health was restored sufficiently for him to contemplate a walking tour, for exercise and fresh air were always his best medicine. Across the Niagara River, so close that he could see the people who lived there, lay the United States of America whose experiment in democracy had fascinated him for as long as he could remember. Crossing the river at Lewiston, he viewed the Falls from the American side, then walked through the magnificent forested hills and rolling country along the Genesee River. This was the state that had elected Governor DeWitt Clinton who two months before had broken ground for the Erie Canal. Here was no governor who could prorogue parliament in a huff. Everywhere Gourlay felt the vigour of a purposeful people, so different from the constant complaints in the Niagara District. Americans were selling wild land for $1.25 an acre, a price that put a million and a half dollars yearly into government coffers. This country had been opened up ten years after Upper Canada and was already ten years ahead of it. In 1790 its population was 960. By 1810 it was 229,148. There were capital turnpike roads here, the equal of any in England. With every step he took, Gourlay's spirits rose and by the time he reached Auburn he had reached certain conclusions. First, it was self-evident that North American colonies could not be retained by Britain for many years on principles less free and independent than those that governed the adjoining country. Second, he would become a land agent in Upper Canada and cross the Atlantic every year to select immigrants.[3] Third, he would solicit Administrator Samuel Smith's approval of his plan for a statistical account of the province, apply for land, and by Christmas be home to settle his affairs at Deptford Farm. Off he set along an excellent American road. He could show Upper Canada how to build one as good!

In high spirits he returned to Queenston where he found letters from his wife. "The Duke's new solicitor has written to Nicholson [their lawyer] that if a reference is to be made, you must deposit £1800 that you will abide by the award! Mr. N. considered this an insult and made no reply to the letter." Then quite innocently his wife mentioned that his cousin, Mr. Wilson, had just died in London.[4] Now the matter of the lottery money would be revealed and Wilson's estate would demand its money. Pushing the troublesome matter from his mind, he answered her letter, told her to inform her brother Thomas that he would never declare bankruptcy, wrote a note to each of his children, and sat down to polish the question-

---

[3]*Ibid.*, #17, p. 180.
[4]Deptford Farm, Sept. 11, 1817, *ibid.*, #25, p. 310.

naire that he was planning to publish in the *Upper Canada Gazette* so that township officials could answer it.

This is the text of the questionnaire:

1. Name, Situation, and Extent of your Township?
2. Date of the first settlement of your Township, number of people and inhabited houses?
3. Number of Churches or Meeting Houses; number of professional Preachers and of what Sects?
4. Number of Medical Practitioners?
5. Number of Schools, and the fees per quarter?
6. Number of Stores?
7. Number of Taverns?
8. Number of Mills, and of what description? with the rate of grinding, sawing and carding wool?
9. The general character of the Soil and Surface?
10. The kinds of Timber produced, naming them in order, as they most abound?
11. What Minerals, if any, have been discovered or indicated; Coal, Limestone, Iron, Stone, Plaister of Paris, Salt Rock, Salt or other remarkable Springs?
12. Building Stones, if any, of what quality, and how much per Toise they can be obtained for at the Quarry?
13. If Bricks have been made, and their cost per Thousand?
14. If Lime is burned, and the price per Bushel at the Kiln?
15. Wages of Blacksmiths, Masons, and Carpenters; and the rate of their piece-work respectively?
16. Wages of Common Labourers per annum, per Winter month, per Summer month, per day in Harvest; also, wages of Woman Servants per week, for Housework, for Spinning?
17. Price of mowing grass for Hay: price of Reaping and Cradling Wheat; saying in each case if board and lodging is included?
18. Cost of Clearing and Fencing a given quantity of Wood Land; say five Acres, by contract?
19. Price of a good Work Horse four years old: also, a good Cow, Ox, Sheep, of the same age?
20. Average quantity of Wool yielded by Sheep: and what price the Wool now brings per Pound?
21. Ordinary time of turning out Beasts to Pasture, and of taking them Home into the Yard or Stable?
22. Ordinary endurance of the Sleighing Season, and of commencing Ploughing in the Spring?
23. Ordinary Season of Sowing and Reaping Wheat?

24. Quantity of Wheat required to Sow an Acre, and how many Bushels per Acre are considered an average Crop?
25. Quality of Pasture: 1st. as it respects Feeding, and what Weight an Ox of four years old will gain with a Summer's run: 2nd. as it respects Milk and quality of Dairy Produce, noting the price which Butter and Cheese made in the Township will now fetch.
26. Ordinary Course of Cropping upon new Lands, and afterwards when broken up from Grass: stating also when and for what Crops Manure is applied?
27. If any is let on Shares: to what Extent this is practised: and what are the ordinary Terms?
28. The Price of Wild Land at the first Settlement in the Township: its progressive Rise and present Price: also of Land so far cleared: stating circumstances as to Buildings, Proportion cleared, or Peculiarity, if any, of Local Situation:; referring in every instance to actual Sales?
29. Quality of Land now for sale?
30. State of Public Roads, and if capable of much improvement at a moderate Expense; also, if any Water Conveyance; or if this could be obtained, extended, or improved, by means of Canals, Locks, &c., &c.?
31. What, in your opinion, retards the Improvement, of your Township in particular, or the Province in general; and what would most contribute to the Same?[5]

The first thirty questions would provide him with the statistics he required. The thirty-first, that had a parallel in Sir John Sinclair's concluding query to the parish ministers who answered his questionnaire, "Means by which their situation could be meliorated", would render the province more attractive to immigrants if the answers were heeded.

Then he composed an "Address to the Resident Landowners of Upper Canada" to accompany the questionnaire as an explanation of his purpose.

"Gentlemen," he wrote, "I am a British farmer, and have visited this province to ascertain what advantages it possesses in an agricultural point of view. After three months residence I am convinced that these are great,—far superior indeed to what the mother country has ever held out, either as they concern speculative purchase, or the profits of present occupation. . . . Gentlemen . . . you should make known the state of this country; you should advertise the excellence of the raw material which Nature has lavishly spread before you; you should

---

[5]*Statistical Account I*, pp. 270-4.

inspire confidence, and tempt able adventurers from home. At this time there are thousands of British farmers sickened with disappointed hopes, who would readily come to Canada, did they but know the truth: many of these could still command a few thousand pounds to begin here; while others, less able in means, have yet preserved their character for skill and probity, to entitle them to the confidence of capitalists at home, for whom they could act as agents for adventure."

He said that the four or five thousand settlers who had come that year were regarded in Upper Canada as a great accession to the provinces, but England would benefit from sending fifty thousand annually.

"The present emigration from England affords no relief whatever to the calamity occasioned by the poor-laws. Thousands and tens of thousands of paupers could be spared, who cannot possibly now get off for want of means, but who would be brought over by men of capital, were confidence for adventure here once established.

"The extent of calamity already occasioned by the system of poor-laws, cannot be even imagined by strangers. They may form some idea, however, when I tell them, that last winter I saw in one parish (Blackwall, within five miles of London), several hundreds of able-bodied men, harnessed and yoked, fourteen together, in carts, hauling gravel for the repair of highways; each fourteen men performing just about as much work as an old horse led by a boy could accomplish."[6]

If Britishers were encouraged to come to Upper Canada, he continued, the recent government restrictions on American immigration would not matter. The late war had shown how strong was the tie between Canada and Britain. If Canada were self-sustaining, British taxpayers would no longer have to spend millions on its defence and government, but would receive a proper return in increased commerce. In order to achieve that desirable effect, the first step should be the publishing of an authentic statistical account of Upper Canada in which every resident land-owner would have a hand. He intended his questionnaire to be the basis for that publicity.

When his questionnaire and Address were finished, Gourlay was ready to leave for York. Dickson accompanied him to the dock, for he considered the project would be a spur to immigration. As they waited for a stiff wind to subside so that the *Frontenac* could weigh anchor, Dickson read the first part of the Address. He was moved to caution Gourlay about a statement he had made at the beginning:

---

[6]*Upper Canada Gazette*, Oct. 30, 1817; *Statistical Account I*, clxxxvi. It may be noted here that at this time many Britons regarded emigration as exile and required encouragement to allay their fear.

London Corresponding Society,

Held at the Bell, Exeter-Street, Strand.

MAN as an Individual is entitled to Liberty—it is his Birth-right.

As a Member of Society, the Prefervation of that Liberty becomes his indifpenfable Duty.

When he affociated, he gave up certain Rights, in order to fecure the Poffeffion of the remainder;

But, he voluntarily yielded up only as much as was neceffary for the common Good:

He ftill preferved a Right of fharing in the Government of his Country;—without it, no Man can with Truth call himfelf FREE.

Fraud or Force, fanctioned by Cuftom, with-holds that Right from (by far) the greater Number of Inhabitants of this Country.

The few with whom the Right of Election and Reprefentation remains, abufe it, and the ftrong Temptations held out to Electors, fufficiently prove that the Reprefentatives of this Country feldom procure a Seat in Parliament, from the *unbought* Suffrages of a Free People.

The Nation at length perceives it, and teftifies an ardent Defire of remedying the Evil.

The only Difficulty, therefore, at prefent is, the afcertaining the true Method of proceeding.

To this end, different and numerous Societies have been formed in various Parts of the Nation.

Several likewife have arifen in the Metropolis, and among them, (though as yet in its Infant State) the CORRESPONDING SOCIETY, with Modefty intrudes Itfelf and Opinions, on the Attention of the Public, in the following Refolutions:

Refolved,—That every Individual has a Right to Share in the Government of that Society of which he is a Member—unlefs incapacitated:

Refolved,—That nothing but Non-age, Privation of Reafon, or an Offence againft the general Rules of Society, can incapacitate him.

Refolved,—That it is no lefs the RIGHT than the DUTY of every Citizen, to keep a watchful Eye on the Government of his Country; that the Laws, by being multiplied, do not degenerate into *Oppreffion*; and that thofe who are entrufted with the Government, do not fubftitute *Private Intereft* for *Public Advantage.*

Refolved,—That the People of Great Britain are not *effectually* reprefented in Parliament.

Refolved,—That in Confequence of a *partial, unequal,* and therefore *inadequate Reprefentation,* together with the *corrupt* Method in which Reprefentatives are elected; *oppreffive Taxes, unjuft Laws, reftrictions of Liberty,* and *wafting of the Public Money,* have enfued.

Refolved,—That the only Remedy to thofe Evils is a fair, equal, and impartial Reprefentation of the People in Parliament.

Refolved,—That a fair, equal, and impartial Reprefentation can never take Place, until all *partial Privileges* are abolifhed.

Refolved,—That this Society do exprefs their *Abhorrence* of Tumult and Violence, and that, as they aim at Reform, not Anarchy, Reafon, Firmnefs, and Unanimity are the only Arms they themfelves will employ, or perfuade their Fellow-Citizens to exert, againft ABUSE OF POWER.

Ordered,—That the Secretary of this Society do tranfmit a Copy of the above to the Societies for Conftitutional Information, eftablifhed in *London, Sheffield,* and *Manchefter.*

By Order of the Committee,

April 2, 1792.

T. HARDY, Secretary.

Resolutions of the London Corresponding Society
drafted by Maurice Margarot.

Oliver Gourlay, 1740-1819, Oil, c. 1790.

Mrs. Robert Gourlay, Pastel, c. 1809.

Mrs. Robert Gourlay, c. 1780-1820, by Sir David Wilkie, c. 1814.

Robert Gourlay, 1778-1863, by Sir David Wilkie, c. 1814.

Robert Gourlay, c. 1860.

Robert Gourlay, 1863,
reproduced by permission of Ontario Historical Society
from *Papers and Records,* 1917.

Reproduction of a portion of a letter from Robert Gourlay to his daughter Jane, Ohio, May 28, 1836.

"When I speak in this sanguine manner of the capabilities of Canada, I take it for granted that certain political restraints to improvement will be speedily removed. The able resolutions brought forward at the close of your last session of parliament, and the opinion of every sensible man with whom I have conversed on the subject, gives assurance of this. My present Address, therefore, waves all regard to political arrangements: it has in view, simply to open a correspondence between you and your fellow-subjects at home, where the utmost ignorance prevails with respect to the natural resources of this fine country."

The resolutions that Gourlay praised so carelessly had been the cause of the dissolution of parliament a few months before. The *Upper Canada Gazette* had been forbidden to publish them, and their existence was known throughout the province only by word of mouth. Dickson warned Gourlay that he was treading on thin ice, but caution was not Gourlay's strong point. He took back his Address before Dickson could read further.[7]

The two men rejoined Mr. M'Donnell, former Speaker of the assembly, who had come down to the dock with them. When the wind abated, Gourlay proceeded confidently towards York where he intended to show his Address and questionnaire to Administrator Smith, to Chief Justice William Dummer Powell, and to the surveyor-general, Mr. Ridout. He had no intention of asking the approval of the Reverend Dr. John Strachan whom Dickson had described as "arrogant, but a good-hearted little fellow", for Gourlay did not feel that he wished to make friends with a man who in his opinion had contributed to the massacre of Lord Selkirk's innocent settlers at Red River.[8] The authorization of Colonel Smith and Judge Powell should be sufficient to allow the publication of his questionnaire and Address in the *Upper Canada Gazette*. It was with a light step and confident air that Gourlay disembarked at York to the sound of the cannon shot that announced the arrival of the *Frontenac*.

As he set foot in the capital of Upper Canada, the only thing that impressed him about the little cluster of wooden houses was the regular layout of the streets. He went immediately to the Mansion House, a long, two-storeyed clapboard building. There, in the hotel patronized by members of the legislature during its sessions, he renewed his acquaintance with two of his recent shipboard companions, a navy captain who had just been granted 1,200 acres and a clergyman who had been given 600 acres. If men with no farming experience whatever were so well-treated, he con-

---

[7]*The Banished Briton*, #17, p. 177. Strachan asserted that Clark and Dickson had helped Nichol prepare these same resolutions. Strachan to Gore, York, May 22, 1817, G. Spragge ed., *The John Strachan Letter Book*, p. 139.
[8]*Ibid.*, #22, p. 260; #20, p. 220.

sidered that he might expect much more, for he was a scienctific agricul-
turist about to prepare an agricultural survey of the province. The oc-
cupant of the Mansion House whom he liked immediately was Mr.
Charles Fothergill, an Englishman who had come to York with his family a
few months before. Gourlay and Fothergill had much in common. Both
men were tall, dark, handsome, well-educated, with pedigrees as long
as their arms. Fothergill had just been granted twelve hundred acres half
way between York and Kingston. Surveyor-general Ridout had held up
several other grants in the vicinity while Fothergill decided which location
pleased him most. Once again Gourlay reflected that he should be given
preferred treatment. Fothergill was a horseman and ornithologist but he
was not a scientific farmer.[9]

The next day Gourlay set off to see Colonel Smith with his questionnaire, his Address and his letters of introduction from legislative councillors Dickson and Clark. Smith showed great interest in Gourlay's enthusiastic presentation of his proposed publicity for the province by means of a
statistical account. He was so accommodating that Gourlay was glad he had
ignored the advice of his Niagara relatives who had said he would waste his
time going to York to petition for land.[10] Colonel Smith shook Gourlay's
hand warmly and invited him to visit him at his summer home a few miles
westward along the lake near Etobicoke.[11]

In rising spirits Gourlay swung down the street to visit another of the
chief dignitaries of the little capital. Chief Justice Powell was a loyalist who
maintained a foothold in both Canada and United States. He had not
approved Lieutenant Governor Gore's proclamation against American immigration, and he disliked Gore's absolute rule which even countenanced
arbitrary imprisonment in peace-time. Chief Justice Powell readily consented to Gourlay's request to use the columns of the government *Gazette*
to publicize his project.

All the leading men of York seemed to have praise for Gourlay's project—Administrator Smith, Chief Justice Powell, Attorney-general D'Arcy
Boulton and his three sons, William Jarvis, the provincial secretary,
Colonel Cameron, Colonel Smith's secretary, Colonel Wells, Captain
Fitzgibbon, and Mr. Ridout, the surveyor-general, head of one of the
busiest departments of the government.[12] Mr. Ridout related some of the
difficulties of his work. Several years before he had reported them to the
Colonial Office but the Colonial Office had ignored his predicament. He
offered to show Gourlay the report if Colonel Smith would allow him.

9J. Baillie, "*Charles Fothergill, 1782-1840*", CHR, Dec. 1944, p. 380. Fothergill's ambition
   was to write a natural history of the British empire—impressive but unachieved.
10*The Banished Briton*, #19, p. 209.
11*Ibid.*, #17, p. 180.
12*Ibid.*, #22, p. 240.

He would also be glad to show him the maps and records Gourlay was asking to examine, if Colonel Smith would give a written order.

Gourlay then went to request a land grant from John Small, the clerk of the executive council and of the land council. Small filled both positions because the members of both bodies were practically the same. Mr. Small nodded and said that the land council would consider the matter. Gourlay was not a man to sit idle while waiting for a decision. He walked to the printing office to supervise the printing of his Address and questionnaire. The compositor in the print shop of the *Upper Canada Gazette* read the Address before he set it in type.[13] He, like Dickson, advised Gourlay against antagonizing certain members of the government by the reference to the "able resolutions" put forward by the assembly. He also indicated that some people might be afraid to sign their names to the township reports as Gourlay had requested. Gourlay was in a mellow mood. He changed the reference to the resolutions to "Growing necessity" and added that the reports might be submitted unsigned. Then he hired a horse so that he could look around while he was waiting.

Northward he rode up Yonge Street, past the dump where York butchers threw the offal of their trade, and past five miles of land owned by York officials who used it as a source of firewood and speculation. About eight miles north of the village he came to a building of singular appearance. It resembled "one of our English staddlebarns dropped from its pillars. Its measurements gave seventeen paces by nine, and out of the roof shot a little stove chimney of brick. A person standing near informed me that it was a Church, and one of the Church of England. I forthwith inquired who was the Clergyman: there was none, specially appointed for this place of worship; but Dr. Strachan did duty in it, once a month. What! said I, has this building been erected for so little benefit to the country! Are there no Presbyterians, Methodists, or Baptists who could occupy it the three vacant Sundays of the month?—'The doctor, Sir,' replied my informant, 'will let nobody preach here but himself.'" Gourlay was astonished at this selfish spirit in a man of God. In the Niagara peninsula, the first churches were used by several denominations freely.

As he rode back to York, he reflected on the spirit that ruled this man, this "passion which had characterized power—lifted up priests, through all the ages of the heirarchy . . . this spirit of intolerance," the very opposite, indeed, of the opinions that he had heard expressed by some of the higher church dignitaries in England. He had dined one day with the Bishop of Llandaff who declared that he would receive any man of any sect who would acknowledge Christ and a future state of rewards and punishments.[14]

13*Ibid.*, #17, p. 177.
14*Ibid.*, #22, p. 238.

Almost to the day when Robert Gourlay leapt down from his horse to measure the little church overlooking Hogg's Hollow and to muse on the rector's lack of tolerance, Dr. Strachan was describing the progress of his work in York to his superior, Bishop Mountain in Quebec. "In this neighbourhood the Methodists are losing ground at present. A church has been built on Yonge St. where I preach once a month to their great annoyance."[15] His letter erred a little. The Methodists were gaining ground all over the province. Kingston itself had two Methodist ministers to every one of three other denominations. There was not one Anglican minister in Upper Canada outside the comfortable rectories in Cornwall, Kingston, Niagara and York, and not one Anglican adherent west of Niagara. The Reverend Dr. Strachan also told his bishop, "The congregation is numerous when I preach." It could not well be otherwise since he excluded other ministers fiercely from his little satellite.

Gourlay stood looking at the church in the middle of the woods from whose purlieus the settlers gathered on the one day that Dr. Strachan allowed. He marvelled that the rector who carried so much weight in York believed in a closed, established church. Mounting his horse, he rode back through the darkening autumn woods to York, glad that he had not shown his Address and questionnaire to the rector.

In a little town no stranger can walk unobserved. Naturally by now Dr. Strachan knew that a relative of legislative councillors Dickson and Clark was in York asking for land, but the tall, distinguished-looking man who went so easily in and out of the offices of other leading citizens of York did not come to him. Something was going on about which he had not been consulted and he was not accustomed to being treated like a clairt at a sheep-shearing. Unable to restrain his curiosity any longer, he ducked into the printer's office to see what was going on. What he found and his opinion of it he described in a letter to Colonel Harvey, secretary to the governor-in-chief in Quebec and applicant in York for a grant of land in the vicinity of Perth.

The letter said that a Mr. Gourlay had come to York with letters of introduction from Mr. Clark and Mr. Dickson, two land speculators in the Niagara District, who stated that their cousin would be a valuable settler if he could be prevailed upon to remain. Both Smith and Powell had given him every attention.

> "While the Printer was preparing his address I happened to go into the Office and he handed me the proof of I believe the first column I cast my eye upon it, and disapproving exceedingly of the sentiments and general tone and Spirit of the whole I advised the Printer to be

---

15Strachan to Mountain, York, Nov. 18, 1817, G. Spragge, ed. *The John Strachan Letter Book*, p. 144.

on his guard for that such a Paper was highly improper for his Gazette. He replied that the Administrator approved of it and like-wise the Chief Justice. I mentioned to Colonel Smith my opinion of this paper when I had seen the whole, and my opinion that the man was a dangerous incendiary and his Scheme of a topographical work a mere pretence to conceal his real views. My opinions were treated with ridicule by the Chief Justice and therefore made no im-pression on the Administrator."[16]

Unaware that the man whom he had ignored was calling his plan sub-versive, Gourlay was busy with the details of launching it. He had ascer-tained that printed letters could be sent free of postage to every part of the province. Immediately he decided to have 800 copies of his work struck off and mailed to every township official in the province. This would give him more thorough distribution than the limited circulation of the *Gazette*.[17]

He also put into writing his request for land to Mr. Small and to the executive council, for Small now told him that he must appear in person before that body in order to procure a grant of land. He went to see Colonel Smith again. Smith made a memorandum with his pencil and promised to do what he could as soon as he could. Gourlay made it plain that as soon as he learned what quantity of land he would be granted, he would take the required oath of loyalty and pay the fees. He composed himself to wait for an answer. He was just about to set off to spend the weekend with Colonel Smith at his summer home in Etobicoke when the gun was heard booming the arrival of the boat from Kingston. Greatly to his surprise, a message was brought to him that his brother Thomas was among those who had disembarked. Down he rushed to the dock to meet him.

Thomas Gourlay, a quiet young man aged twenty-three, had been apprenticed to a Writer to the Signet in Edinburgh and had expected to farm Glentarkie. His father's bankruptcy blasted the latter hope. When Robert's letters praising the new land arrived at Craigrothie, his father persuaded Thomas to give up his apprenticeship at law and go to Canada to seek his fortune. The two brothers stayed a few days in York while Robert supervised the mailing of the questionnaires, and then they em-barked for Queenston to introduce Thomas to the relatives there.[18]

As they crossed the lake back to Niagara, Robert Gourlay discussed immigration with his fellow travellers. His opinion was that pioneering need not degrade one's character. If the art of settlement were practised

---

[16]Strachan to Col. Harvey, York, June 22, 1818, *ibid.*, p. 163.
[17]*The Banished Briton*, #17, p. 176.
[18]*Statistical Account II*, p. 419.

properly both the man and the land would be improved. It was then that he learned "from unquestionable authority", which he later named as one of the Boultons, that Dr. Strachan disapproved his Address and had said that a stranger like Gourlay had the utmost presumption to make the proposals he had.[19]

The displeasure of one person out of the total population of Upper Canada seemed a small factor in a chorus of approvals, but for the third time in his life, with unerring accuracy, Gourlay had offended some one at the top. Executive councillor Strachan knew what would be some of the answers to the thirty-first question on the questionnaire: "What, in your opinion, retards the improvement of your township in particular, or the province in general, and what would most contribute to the same?" The assembly itself had been dissolved because it had tried to discuss the state of the province and Strachan had approved Gore's prorogation. Now this stranger presumed to re-open the issues. The reverend doctor had already made up his mind. This cuckoo from abroad, this reinforcement for the dissidents in the assembly, this dangerous revolutionary, this cousin of the Niagara land speculators to whom Joseph Brant had sold land,[20] this man who praised American progress in the Genesee district, this man who was proposing to write the same type of book that he himself had in mind,[21] this Robert Gourlay must be strongly opposed for he was a danger to Upper Canada.

---

[19]*Statistical Account I*, p. cccxiv; *The Banished Briton*, #17, p. 176.
[20]In *A Visit to the Province of Upper Canada in 1819*, purportedly written by his brother, Strachan described Brant as that "miserable man . . . of savage ferocity". He was gentler with the Indians when their souls were the issue, not their land. J. Strachan, "Religious State of the Indians in Upper Canada, May 7, 1838; *Statistical Account I*, p. cciv.

13

The Township Meetings

When he returned to Niagara, Robert Gourlay was pleased to see that the *Niagara Spectator* had copied his questionnaire and Address from the columns of the *Upper Canada Gazette* and he assumed that the *Kingston Gazette* had done the same. On November 10, 1817, he sat down to write his thanks to the editor of the *Spectator*. Then, because he never allowed any trifle to go unnoticed for fear that it might grow into a larger issue, he mentioned that he had been told during his return boat trip from York that someone in York did not approve his project.

> "Were it a matter of private concern, it would be much beneath me to notice the whisperings of any one; but under the present circumstances my duty is to speak out, and once for all, to guard the public against murmurs which may retard the promotion of the best interests of the Province. . . . In offering my services to this Province, as Compiler of its Statistics, I offer much labor with small prospect of emolument; and if there is, in York or elsewhere, any one willing to undergo the drudgery, in whom greater confidence can be placed, most gladly will I resign to him the task; which indeed I cannot perform unless *immediately and frankly* assisted in the manner proposed.
>
> "That I am a stranger is not in every sense correct. My near connections were among the first settlers of Upper Canada, and I am proud to say, rank also as its greatest benefactors. Indeed, where the memory of the Hon. Robert Hamilton is so much revered, it is natural for me to claim that country as my home."[1]

Then he drew attention to the parts of the Address that might possibly be the offending portions—his criticism of government failure to tax wild lands, and the rejection of the assembly's resolutions that he regarded as the encroachment of arbitrary power. He concluded his lengthy letter

---

[1]*Niagara Spectator*, Nov. 10, 1817; *The Banished Briton*, #17, p. 176.

to the *Niagara Spectator* with a thirty-second question: "Can you recommend any person peculiarly qualified for arranging and publishing the Statistics of Upper Canada, or are you willing to repose confidence in your humble servant, ROBERT GOURLAY?"

Still no word came from York regarding the land grant he had requested. In the eyes of the land council a thousand hours were as a day, but in Gourlay's opinion the two weeks he had spent in Little York should have been ample for accomplishing his purpose. Now he had cooled his heels another ten days in Queenston. He was in a hurry for an answer so that he could return to England to settle his affairs. If he must wait longer he decided to travel 100 miles to Dereham Township to inspect his property there. His project for compiling his statistical account would not suffer during his absence for his Niagara friends had endorsed it in the *Niagara Spectator*. The regular township meetings were not scheduled to be held till January but dates for the first statistical account meetings had already been set by a group of leading Niagara magistrates.[2] Stamford Township convened the first meetings on November 17, probably the day on which Robert and his brother Thomas set out on their trip into the western part of Upper Canada.

Content in the knowledge that the project was self-propelled, the two brothers set off along the road that skirted Lake Ontario through land known by the names of the townships that had been laid out roughly nine miles at the front and twelve miles deep. Through Niagara Township they went, through Grantham, Louth, Clinton and Grimsby. Over halfway to the Head of the Lake they paused for the night in Saltfleet Township. En route they met the clergyman who said he had recently received 600 acres. From him they learned that, contrary to Small's statement to Gourlay that all applicants for land must appear before the land council, the clergyman had taken the oaths and been assigned his land without ever laying eyes on the council.[3] If that were the truth, Gourlay decided that Thomas should return to York, take the oaths and receive his land immediately for he did not have to return to Britain to prepare a family for emigration. Robert sat down to write two letters for Thomas to take to York. The first he addressed to Small, the clerk of the executive council, asking him if his petition had been presented to the council yet. The second was to Colonel Cameron, Colonel Smith's secretary, requesting a list of deserters during the recent war, how many returned to the province after the war, how many were tried for treason, condemned or acquitted.[4]

2*Ibid.*, p. 178. The township reports are found in *Statistical Account I*, pp. 269-625.
3*The Banished Briton*, #19, p. 210.
4Saltfleet, Nov. 18, 1817, *ibid.*, #19.

Why did he wish information on one of the touchiest subjects in the province? Once again Gourlay was talking to all kinds of people to find out how they were living. There were many ex-Americans in the Niagara District but as he conversed with them he found no evidence of the disloyalty commonly attributed to them by members of official circles. He was also talking with men who had suffered war losses but had received no compensation. The government headed by Gore had said that they could be paid out of the sale of lands confiscated from American deserters but Gore's proclamation restricting American immigration had eliminated buyers. These ex-Americans were dissatisfied with the edict but not disloyal, Gourlay concluded.

Next morning the two brothers reached Barton Township, now obliterated by the city of Hamilton. Here Thomas branched east to York and Robert went west from Ancaster along the Dundas Road, named after Henry Dundas, Lord Harry the Ninth of Scotland. Ancaster was the town fifth in importance in the province. Richard Beasley, former partner of Richard Cartwright in Kingston and Robert Hamilton in Niagara, owned a store and mill in Ancaster as well as an estate and mill at Coote's Paradise at the foot of the escarpment. Beasley needed settlers for his Grand River land, and as he had come from Albany himself, he saw no reason why they should not be Americans.[5] Already eighty-seven men in Ancaster were preparing to meet on November 29 to consider Gourlay's questionnaire. They had not taken long to realize the promotion possibilities of his project. James Durand would chair the meeting.

Leaving Ancaster, Gourlay continued along the southern boundary of Dickson's township of Dumfries. The short dark November days brought rain and a threat of snow that chilled the bones. It was a poor time of year for travelling, but nothing could deter Gourlay when his mind was made up, not even the execrable Canadian roads. Now he turned south into Burford Township at Brant's Ford on the Grand River (now the city of Brantford), for on November 27 he wrote back from Burford to Administrator Smith, critical of "the tardy way in which I get through the country", and asking again about the progress of his petition through the council.[6] He also wrote to Mr. Jarvis, the provincial secretary, asking for a copy of the proclamation issued by Lieutenant Governor Simcoe and inquiring if he had the list of deserters ready for him to see on his return. Certainly many members of the administration knew by now that Gourlay was giving more than a cursory glance at the province.

---

[5]Beasley said government land policy favoured loyalists. Chief Justice Elmsley accused him of conspiring to seize Upper Canada for the United States. Lillian Gates, *Land Policies of Upper Canada* (University of Toronto Press, 1968), p. 44.
[6]*The Banished Briton*, #19, p. 207.

Why did he ask to see the proclamation that Simcoe had advertised in the United States in 1792 inviting all to come north for free land? When Simcoe had ordered the surveyors "to cut and slightly mark" a road to London, his plan was to grant land along it to settlers who would promise to make a road through it.[7] Some had fulfilled their commitment, but as soon as Simcoe was recalled, the "land grabbers" in York secured many of the choice lots bordering the Dundas Road. The actual settlers were now in isolated locations, angry at the broken promise that had tempted them so far into the wilderness. Gourlay had seen at Perth that a concentration of settlers was necessary to keep them happy. By now he was determined to include more in his book than a table of answers to his questionnaire. He would print Simcoe's *Proclamation* and some laws of the province as well. The violation of the *Proclamation* was at the root of the growing number of complaints against the government that he was hearing as he penetrated the western section of the province. The farther he went, the higher rose his opinion of the Americans who comprised the bulk of the settlers here.[8] He could see no reason for prohibiting their entry to the country or for making a blanket judgement of disloyalty against them.

Now he was seeing first-hand the curse of the clergy and crown reserves. They were set chequerboard fashion and undeveloped in each township he traversed, two-sevenths of the land, or ninety-six lots of 200 acres each in every township. The Niagara District was free of them because its development had begun before the Constitutional Act had been passed in 1791. One complete township in the Western District had been set aside to compensate for the quota that had been missed in the Niagara District. Captain C. Stuart, a justice of the peace of the Western District, described them: "like rocks in the ocean they glare in the forest unproductive themselves, and a beacon of evil to those who approach them".[9]

From Burford Gourlay travelled diagonally south to the next township, Norwich. Here, in a sizable Quaker settlement with two blacksmiths, he was entertained by Peter Lossing, its leader. Lossing had come from New York state in 1809 to purchase 15,000 acres (from William Willcocks) which he sold at cost to the neighbours he intended to bring up. It was Lossing who guided Gourlay on horseback to his land in Dereham township for there was no road into it and only one settler in the whole township.[10]

---

[7]*Statistical Account II*, p. 311.

[8]*Ibid.*, p. 422.

[9]*Statistical Account I*, p. cccvi.

[10]*Emigration and Settlement on Wild Land*, p. 11. Simcoe set aside the townships of Norwich and Dereham for revenue for schools. When this was advertised in New England papers, and no buyers came forward, Robert Hamilton acquired 6,000 acres in Dereham and 9,000 in Norwich.

This was his land, his land and Jean's and their children's. This was the land he had travelled thousands of miles to see. It lay before him, couchant with fertility beneath the mighty trees, some of the best land in what would be known as the Garden of Ontario. It was in his power to turn this land into productivity just as he had done with Deptford Farm. It was big enough to support him well but it was too small for the poor of England whom he hoped to bring, and if he brought them so far inland, his costs and their hardships would be doubled. Surely he could get a location closer to roads and markets than this. His plans for his immigration scheme had grown with every mile of the vast country he penetrated. He was beginning to think that if the government would give him the management of the public lands in Upper Canada for thirty years, he could maintain two regiments for his majesty, keep in repair all the forts, and for the last twenty years of the term he could pay the British government an annual rent of £100,000 sterling.[11] As matters stood, the lands of the crown were unsaleable and stagnation and discontent were the trademarks of the country. There was little hope that Canada would ever cease being a burden to the British taxpayer, but if his "Great Offer" were accepted, Gourlay could foresee swift prosperity.

He had seen his land, and had he followed his original intention he would have returned to Niagara for the township reports which he was certain would be ready for him now. What he hoped was indeed true. Behind him the meetings were being held, sometimes as many as three a day. They were meeting in every section of the province, in townships and towns: November 17, Stamford; November 18, Nichol; November 21, Wellington Square (now Burlington); November 22, East Flamborough; November 26, Kingston Township and Thorold; November 27, Trafalgar; November 28, Kingston; November 29, Grantham and Ancaster. They were grasping the chance to bring increased prosperity by reporting about Upper Canada's rich but empty land. Still they met: December 1, Canborough and Caister, and West Flamborough and Beverley; December 2, Blenheim and the first concession of Burford together, and Raleigh Township, the first to meet in the western extremity of the province that was later than the Kingston and Niagara Townships in receiving its mail; December 4, Windham and Willoughby; December 5, Burford, Norwich (likely the day before Gourlay's arrival at Lossing's) and Walsingham. Like signal fires on the hills, they had sprung up all around, ready to promote a project so clearly designed to further the well-being of the province.

Gourlay might have turned back after he had seen his land, but he was still the man who in his *Letter to the Earl of Kellie* had evinced a

---

[11]Feb. 1, 1818, *The Banished Briton*, #34, p. 466; *Statistical Account I*, ccccxcviii. As usual Gourlay failed to foresee the reaction his offer would elicit.

compulsive passion for truth in private and public affairs. By now he had received so many complaints about the government and been asked so many questions with regard to regulations emanating or supposed to be emanating from York, all of them increasing as he travelled westward, that he "began to suspect that all my fond hopes of promoting the welfare of the Province, by the publication at home of my Statistical Reports, would prove in vain. Who will come to a country, thought I, where public business is neglected, government faith held so light, and the rights of property trifled with?"[12] He decided to go on to the western extremity of the province to assess conditions all the way.

He was now at the border of the Talbot settlement, the earliest in western Upper Canada except for the trading posts. The Talbot settlement had its beginning in 1803 when Colonel Thomas Talbot, former private secretary to Colonel Simcoe, returned to Canada with permission from the Colonial Office to settle in Dunwich Township. Robert Gourlay had reached the domain of the man who had been authorized by the British government to claim 200 acres for every settler he located on fifty of his original officer's grant of 5,000 acres. Talbot had not confined himself to his original allowance. By 1817 he had infiltrated five townships—Dunwich, Bayham, Malahide, Yarmouth and Southwold—an estimated 518,000 acres.[13] In spite of the fact that his holdings amounted to a glut, Talbot too was dissatisfied with York. He hated their fee-splitting ways and favouritism. He saw no reason why he should do the work of settling his people and some clerk in York be paid for it, and why his settlers should have to travel the long distance to York for their land patents. York at this moment was fighting to end Talbot's suzerainty. With settlers arriving daily for land, the land council had to know what land was available. No one knew what Talbot was doing except himself. When General Brock promised some land to two officials, the land office found that Talbot had already bestowed it on settlers of his own choice.[14] They had to swallow this bitter pill and satisfy Brock's favourites with double the amount of land elsewhere. Said Talbot to York, "I'll be damned if you get one foot of land here."[15] Talbot was not at home when Robert Gourlay reached his settlement. He was in London lobbying the Colonial Office over the heads of York's officials, just as Clark and Dickson had advised Gourlay to do.

If Talbot accused the provincial government of greed and favouritism, he himself had ways that were far more peculiar than anything that took

---

12*The Banished Briton*, #16, p. 155.
13G. M. Craig, *Upper Canada: The Formative Years* (McClelland and Stewart, 1963) p. 144.
14F. C. Hamil, *Lake Erie Baron*, (MacMillan, 1955), p. 84.
15N. MacDonald, *Canada, 1763-1841; Immigration and Settlement* (Longmans, 1939), p. 138.

place in the capital. He kept a map of the township lots on the wall of his house at Port Talbot and interviewed prospective applicants through a sliding window that he slammed shut if the applicant displeased him. If a settler were tardy in completing his settlement duties, Talbot rubbed his name off the map, separating the man in one second from any laborious improvements he had made on his land. On the occasions when he materialized out of the wilderness on the streets of York where his peers tried to call him to account during Gore's absence, he silenced them by saying that he had Lieutenant Governor Gore's verbal consent for his actions. It was probably true, for Talbot had turned on his re-membered aristocratic charm on the occasion of Gore's visit to his land in 1806.

Before he departed in the spring of 1817, Gore had attempted to re-plenish the empty treasury by asking Talbot to remit settlers' fees whe-ther settlement duties were complete or not and to spur the rest to take out their patents.[16] York officials claimed that Talbot owed them at least £1,200 in land fees. Talbot told them to go to blazes, for he had not received one penny compensation for his great losses during the war.

Talbot's rule behind the sliding door was autocratic but he had made the long Talbot Road. His settlement was beginning to prosper and even show gratitude to the man who had nursed it into being. No English duke setting out plate for three or four hundred guests on sheep-shearing day had been prouder than he the previous spring when his settlers had held the first Talbot Day in honour of the anniversary of his coming to Dun-wich. This was the man whose admittance of American settlers York officials were seeking to curtail, and whose unregulated expansion York was trying to check by offering him 4,200 final acres.[17] Talbot would return from London with permission to admit Americans to his land, to allow his settlers to fulfil their settlement duties before they paid their fees, and to issue patents to his settlers himself without their making the arduous journey to York. Thomas Talbot was as baronial as ever.

Now Gourlay was in the Talbot settlement at Yarmouth, in the inn of Dr. Lee, an ex-American. Dr. Lee was so impressed with Gourlay's project that he would accept no payment for his food and lodging.[18] Everyone in this district was welcoming the man who had written the Ad-dress and circulated the questionnaire. Most of the meetings were held before his arrival. It was December 7 when he knocked on Colonel Mahlon Burwell's door only to find him absent. Burwell was leader of the settle-

[16]G. Paterson, *Land Settlement in Upper Canada* (16th Report, OA, 1921), p. 123; C. O. Ermatinger, *The Talbot Régime* p. 91, says £4,000 in fees and survey fees. In 1819 Talbot was selling land for 10s an acre, F. C. Hamil, *Lake Erie Baron*, p. 96.

[17]Talbot's friend in London was Whig Lord Dacre who thought he too could "revive Arcadia" by founding a colony in Canada "where all the errors of the old could be avoided". Powell Papers, TPL, A 27.

[18]*Statistical Account II*, p. 454; *The Banished Briton*, #16, p. 155; #22, p. 240.

ment during Talbot's absence. The Yarmouth meeting was held in Burwell's home on December 10. It is likely that the meeting was arranged
before Gourlay appeared. The earliest meeting in the western part of the
province was held on December 2 in Raleigh Township, five townships to
the west of the Talbot settlement. At Bayham twenty-two men met in
John Lodor's house to affirm that they had seen Gourlay's Address in the
Upper Canada Gazette and approved the stimulus that his statistical
account would give to immigration from Britain. Some of these reports
were signed by every man present, so anxious were they to be part of the
large scheme. It was pleasant for Gourlay to be a cynosure instead of being
ignored as he had been when he had issued his call to the farmers of
Great Britain. "In several places I found . . . respectable meetings gathered
together and actually at work on the business. Seeing things in such a
train, I could not resist an inclination to do my utmost for people so
willing to help themselves."[19]

From Southwold he turned north to the Longwoods road again.[20]
Past Dunwich and Aldborough he travelled to Orford which contained
the settlements of the Moravian Brethren. Here was the site of the battle of
Moraviantown where Tecumseh was killed. Gourlay stayed long enough to
investigate the stories of the death of that brave Indian chief, and satisfied himself that it had been American Colonel Johnson who was responsible for killing him.[21] He was pleased to find that the Moravian Brethren
were vaccinating the Indians against smallpox. On he went through
Howard, Harwich and Raleigh, through Tilbury East and West, Rochester
and Maidstone to Sandwich on the Detroit River, the uttermost end of
his journey. In Sandwich he was entertained by John Askin, one of
the leading fur trading merchants, friend of Hamilton, Clark and Dickson.
The meeting here took place on December 17, whether before or after his
arrival cannot be ascertained.

On December 20 he was ready to return. In this vast country, the length
of his journey from Niagara to Sandwich and back was longer than the
distance from Land's End to John o' Groats, five or six hard-driving days
in good winter travelling weather each way. By December 26 Gourlay was
at Putnam's in Dorchester Township. There he sent a letter back with the
postman to Sandwich asking Mr. Askin if he could procure him a Detroit
publication by a Reverend Mr. Richard on the rise and fall of the waters
of the St. Lawrence. The scope of his book was expanding daily in his
mind, and with the township meetings being held all around him, he
hoped that he would have all the reports in his hands before he returned
to England. He traversed the rest of the western snow-belt in zero weather

---

19*Ibid.*, #19, p. 210. Timing is important here, for the snap judgment has been that
   Gourlay fomented discontent in western Upper Canada.
20*Statistical Account I*, p. cccv.
21*The Banished Briton*, #28, p. 392.

and was back in Queenston in the first week of January after an absence of seven weeks.

His first visit was to the post office. There he found that the letter he had posted to his wife weeks ago from the Western District telling her that his departure was being delayed had arrived in the post office only a few minutes before he did, instead of being on its way to Britain. The Upper Canadian postal service was not Mercury-footed. He let everyone know his opinion for he was wild at the thought of his wife's anxiety as she waited daily for his return.

There were letters from her, the first one dated October 19, 1817. "Well, my dear Gourlay, we have had another scene of distraint here! . . . Lamb and a sheriff's officer called yesterday afternoon and demanded the rent. They were told that Mr. Nicholson would settle it as soon as he and Mr. Barton had ascertained to what it amounted. . . . John [the head ploughman] told me last night, after many clearings of his throat, that he had £60, which he hoped I would make use of if I needed it! . . . Yesterday was our harvest-home, and we had a very merry set in the lower part of the house; but some of our household seem to feel for us very much. I was struck with Eliza's excessive paleness and asked if anything was the matter. 'No, Ma'am,' she answered, 'but these people (meaning Lamb and his law-hounds) surprised me a little;' then she burst out crying, and ran out of the room. The children, poor thoughtless little things, have suffered nothing; and you must not let it vex you, my dearest Gourlay . . ."[22]

The next letter bore the news that she had managed to meet the rent with her mother's help and that of one of Robert's brothers-in-law had bought Glentarkie from his father's bankrupt estate at the upset price of £26,000. Another said, "The girls were enchanted with your letters, and they really have been uncommonly attentive ever since to all their tasks."

With loving pride Gourlay answered: "My ever dear wife—. . . As I travelled along, I had the honour of several dinners given me by magistrates. . . . Thus, my dear Jean, I am diverted from brooding on evils at home, which I cannot prevent. Should I call that evil which makes me proud of my wife—more and more proud of her? . . . I shall not now sail from New York till the spring, and it will be May before we meet, &c."[23]

Again he wrote a letter to York about his two-month-old application for land and sent it there by the hand of Samuel Street, partner of Thomas Clark.[24] Surely, Gourlay thought, Street will bring an affirmative answer for a man who was ready to work for the good of the whole province and for the poor of England.

[22]*Ibid.*, #19, p. 210.
[23]*Ibid.*, #25, p. 311.
[24]Niagara Falls, Jan. 5, 1818, *ibid.*, p. 312.

14

Land of Lice

While he waited for a break in the silence that encompassed York, Gourlay read the back numbers of the *Niagara Spectator*. Every issue exploded with letters pro and con his *Address to the Resident Landowners of Upper Canada*. Some said he had been sent by the Prime Minister of England to collect information for taxation purposes. Some said all he wanted was money. One accused him of being a vile democrat. Another scoffed: If Mr. Gourlay were so sorry for the poor of England who were harnessed to ploughs, why didn't he stay home to help them? A letter copied from the American *Geneva Gazette* praised him:

> "Secure in the enjoyment of national and individual blessing our-
> selves, we desire to see similar blessings extended to the whole human
> race. The redundant population of England might find in Canada
> a safe asylum from famine and the sword, to which it is alternately
> exposed; . . . Mr. Gourlay appears to be a man of sound sense, of true
> patriotism, and of correct benevolent views . . . and, if we are not
> much mistaken, he possesses that rare combination of talents, industry
> and judgment, which is calculated to effect good in the world. We
> heartily wish him success in the cause he has undertaken, because
> it is the cause of humanity."[1]

There were also references to the letter Gourlay had written to the *Spectator* just before he left on his western tour regretting that his project had offended an executive councillor whom he did not name. He had omitted the name in order not to pain the particular man involved. Instead, every councillor thought that he was the one. Gourlay wrote another letter explaining that the unfriendly councillor was Dr. Strachan, for both Colonel Smith and Chief Justice Powell had approved his project.[2]

---

[1]*Niagara Spectator*, Dec. 18, 1817; quoted from *Statistical Account II*, App. cxv.
[2]*Niagara Spectator*, Jan. 9, 1818; quoted from *Statistical Account II*, App. cxxviii.

If Dr. Strachan felt that he had been injured in any way, Mr. Gourlay would be glad to answer him for he wished to keep no secrets in public matters. He also published a list of the townships that had not yet reported. It was obvious that not one report had emerged from the Home District of which York was the centre. Two weeks later he published verbatim Simcoe's *Proclamation* regarding admission of American settlers. Neither Colonel Cameron nor Mr. Jarvis had been able to find a copy in York, or so they had written to him. Councillor-cousin Thomas Clark had put his hands on one immediately, for he had not approved Lieutenant Governor Gore's circular that had superseded the *Proclamation*.[3]

That same week the columns of the *Upper Canada Gazette* announced the permanent appointment of Dr. John Strachan to the executive council. With this life appointment the "Teapot Terror" of York was well on his way to being the leading ecclesiastical and political adviser to the lieutenant governor of the province. He had aspired to the position of interim administrator but Colonel Smith had been appointed.[4] He coveted a seat on the legislative council as well as the executive council so that he, like Powell, could listen to the debates of both councils. He had asked the departed lieutenant governor to confer that boon on him. "It was," he would soon write to his friend Major Halton, secretary to the governor-in-chief in Quebec, "the only favour I ever asked of Governor Gore".[5] His permanent residence in York made him constantly available for consultation by the lieutenant governor and for attendance at every session of the executive council. His plans for dominating the assembly were nearing maturity. In May 1817, the month before Gourlay had passed through Cornwall on his way to Niagara, Dr. Strachan had written to the Lord Bishop of Quebec: "In the Lower House I shall by means of my Pupils three of whom are already there possess a growing influence such as no other person can possess . . ."[6] The three pupils, Jonas Jones, Philip Vankoughnett and John MacDonell, were sons of leading conservatives in the loyalist section of the province. Strachan's immediate purpose in writing to the Lord Bishop was to report the progress through the assembly of amendments to the District School Bill of 1807 that would give more control over education to the Anglican church. The long range intention was to make John Strachan as powerful as Cardinal Beaton in his castle beside the sea at St. Andrews or a "Grey Eminence" in Paris.

---

[3]*Ibid.*, p. 445; Clark's copy, p. 470.
[4]Strachan to Harvey, York, June 22, 1818, G. Spragge, ed., *The John Strachan Letter Book*, p. 165.
[5]Strachan to Halton, York, April 10, 1818, *ibid.*, p. 159. Strachan achieved this too in 1820.
[6]Strachan to the Lord Bishop of Quebec (in England), York, May 12, 1817, *ibid.*, p. 129.

Already Strachan was preparing a setting in keeping with his enhanced status. The "Bishop's Palace" on Front Street between York and Simcoe, was the first brick house in the capital. It would make him, he said, the best housed man in the province. As Robert Gourlay walked past it, he compared the little church up Yonge Street with the big brick mansion and noticed acidly that God was housed more humbly than His servant. The Doctor's own brother James, who would visit him the next year, looked at the rectory. "Eh, Jock," he said, "you've a housie like a castle and a wifie like a queen. I hope ye coom by it honest."[7] Then he returned to Aberdeen where his brother had financed him to a book shop which he favoured with orders from the Upper Canada legislature and schools. James was cautioned to send them at prices that would not be too low.[8]

In this year of 1818 John Strachan's plans "for the diffusion of moral and religious knowledge which I have been dreaming upon these fifteen years" seemed near fruition.[9] He was at the height of his physical strength. He was a short handsome man with dark springing hair and a strong cleft chin, but somehow the artist who painted him at this time caught the deprecating attitude in the curve of the eyebrows, a humble-haughty air that seemed to say, "Here is only a dominie who seeks to make his way nowhere." The dominie knew where he was going. What others did not yet understand was the depth of his determination to keep the province free from atheistic Americans and the clergy reserves under the control of the Anglican church.

Dr. Strachan did not find it necessary to acknowledge Gourlay's invitation to a debate on the advantage to the province of a statistical account compiled by a man who implied that a change of government policy was advisable. Though there is no evidence available that Strachan forbade them, no township meetings were being held in the Home District. There was not the slightest chance either that Gourlay would receive the land he wanted. His sponsors were two Niagara "land speculators", mere legislative councillors.

On January 6, 1818, Mr. Small, clerk of the executive council, despatched an answer to Gourlay's request for land. "Sir, . . . when you arrive in the Province, with design to establish yourself in it, as a settler, a location will be made in proportion to the opinion then formed of your means to become a useful settler."[10] Thomas Gourlay was given no answer at all to his application. The only other reason Gourlay was given for the refusal was that his application was written on half a sheet of paper instead of a whole one.[11]

---

[7]J. C. Dent, *Canadian Portrait Gallery*, I, p. 108.
[8]See note 6.
[9]Strachan to Major General Sir George Murray, Sept., 1815, *ibid.*, p. 94.
[10]*The Banished Briton*, #19, p. 208.
[11]*Ibid.*, p. 211.

"Schönberg had nae notation" for Gourlay's whistle at this rebuff.[12] By now he and his large plans were known throughout the province and it was impossible to fade away unobtrusively. On January 21 he blasted back in the columns of the *Niagara Spectator*.[13] He said he regarded the answer as unsatisfactory, the language vague and arbitrary. What did "in proportion to my means" imply? Were his means his money, his children, his servants, or his cattle? He said that the development of Upper Canada and the welfare of thousands of people in Britain would be influenced by this rejection. York remained silent. Its official gobbledegook was final.

Many applications for land have been preserved, but Gourlay's is not one of them. Fothergill asked for a larger grant than was usually made to "mere, ordinary, farmers".[14] It is not known how much Gourlay requested. Fothergill accepted his 1,200 acres gracefully but his plans for development were not deep-seated like Gourlay's. Unfortunately for him, twelve hundred acres was now the maximum grant according to the latest pronouncement made by the British Colonial Office. By the time Gourlay arrived in Upper Canada, three methods of land granting were in disrepute: Simcoe's idea of granting townships to men who would form the nucleus of a landed aristocracy; grants with elastic terms to a man like Talbot who was too remote to control; and a government-aided settlement like the one at Perth whose cost had rocketed as civil and military authorities squabbled. The imperial parliament that made the rules was divided between those who wished to retain the empire and those who objected to its cost. The land councils themselves could be compared with people like Henry Dundas who loved the patronage he could dispense from his position in the Admiralty. As recently as 1816 the bishop of Quebec had been granted 12,000 acres.[15]

Gourlay was incensed at the rejection for more than personal reasons. He knew that the British government had spent millions on a war but would not spend hundreds on a settlement.[16] He was already aware that no one knew whom the land council would treat with largesse and who would be rejected. The unfairness was beginning to rankle. He had listened to people's complaints all the way from Niagara to Sandwich. Now he had one of his own.

He was more positive than ever that he must examine further the

[12]Hugh MacDiarmid, *A Drunk Man Looks at the Thistle* (Edinburgh, 1926), p. 17.

[13]*The Banished Briton*, #19, p. 208.

[14]March 10, 1817, Upper Canada Sundries, DA.

[15]G. Paterson, *Land Settlement in Upper Canada*, p. 119. The land council was called on to solve a variety of ticklish problems, like requesting two men with similar names to appear on the same day for their grants. They refused land to a man who preferred Tom Paine to the Bible. One man slipped past them. He wrote home: "I live in a fine neighbourhood, there is scarcely a Tory in it." *Ibid.*, pp. 73, 17. Some of the complaining was normal annoyance with changing regulations, like the recent toll of 5s 6d cy. as a check to frivolous applications.

[16]*Statistical Account I*, p. 479.

administration of this province. By February 7 he had over sixty answers to his questionnaire.[17] The answers to the now famous thirty-first question about what was retarding the development of individual townships were explicit. The Niagara and St. Lawrence townships wanted canals. The Midland District reports were loud with the yelp of the small businessmen who wanted settlers with capital to sustain them. Trafalgar Township said that the settlers had no cash and the merchants took away their land as payment for their debts. The chiefs of the Six Nations in Haldimand Township complained that they were not allowed to sell their land, and who would rent it when so much other land could be had free? The great Ancaster meeting of eighty-seven men wrote that only the parliament of the United Kingdom could lay the axe to the monumental sloth of the administration in Little York. The districts that had been devastated by war wanted their compensation immediately. The appointed councils had told them there was no money to pay them, but one answer referred openly to the £3,000 "Spoon Bill" for Lieutenant Governor Gore who had done nothing for the country but retard it, in their opinion.[18] Kingston Township mentioned the "great quantities of land in the fronts and public situations that remain unimproved, by being given very injudiciously to persons who do not want to settle on them, and what is most shameful and injurious, no law is made to compel them to make or work any public road; but this is to be done by industrious people, who settle around. Such lands remain like a putrid carcass, an injury and a nuisance to all around: . . ."[19]

The report from empty Nichol Township, drawn up by Thomas Clark and the Reverend Robert Addison without the assistance of the third owner Robert Nichol who was in London with his grievances, was a frank essay on the immigration situation. The two men made no secret of the fact that they had purchased the land from the Indians at four shillings an acre some twenty years before and they now hoped to sell it at two to four dollars an acre. They described it: "The soil of this township is of an excellent quality, as the size and growth of the trees indicate. It is a deep, sandy loam, with abundance of spring brooks in all directions. There are no hills or other very high land; the surface in general level, with a gentle declivity towards the river on both sides. The prevalent timber is maple, elm, beech, white ash, basswood, black ash, and cherry. The bottoms of the brooks are gravel; but no building stones have been noticed, excepting at the river, which in the whole of its course through this township is on a limestone rock."[20] Clark, Addison and Nichol wanted settlers who

17*Statistical Account II*, p. 462.
18*Statistical Account I*, p. 480.
19*Ibid.*
20*Ibid.*, p. 377.

would farm the land, not let it lie useless. They wanted a change in policy in the land granting department at York that kept the best land for favourites, and sent settlers at random to locations far from roads and markets. They said that hatred of York's land policy was the real reason why settlers during the recent war had been so slow to provide the king's troops with provisions and transport. The government had called them disloyal. This report implied that they were disgusted with the prevailing administration.

Kingston Township took a realistic approach to Gourlay's optimistic claim that restraints to improvement would be speedily removed by the British government. "This assertion ill comports with the notice given by the Board of Trade to our merchants, to set a duty on timber from British America. We do express our belief, to think it impossible so much injury to the people of these provinces can be intended, to please a northern despot, or to answer any policy. If such is the intention, we may be assured, more, like Bute and North, are in council who, by their arbitrary measures, lost to us the, now, United States."[21]

Kingston Township report was the only one that was anti-American. The allegedly disloyal American District of London, in contrast to this parochialism, and alone among the Districts, asked that all the reports be published in the three Upper Canada papers. It had concern for the whole province, not one selfish section.

By now all the Niagara townships but Dickson's township had reported, and most of the townships in the Gore, Western, London and Newcastle Districts. Name after name was signed in support of Gourlay's project. The Jacobs, the Moses', and the Benjamins, the Samuels, the Nathaniels and the Asahels, the Absaloms and the Ezekiels, the Georges, Johns, the Williams and the Conradts, all of them eagerly put their signatures to the cause of mutual prosperity and progress. Dr. Strachan might say that Gourlay was deliberately fomenting discontent in Upper Canada but a study of the reports reveals rather than he was only the dynamite to the dam.

Still no reports had come from the townships close to York, and now a rumour reached Gourlay's ears that some of the reports that he had asked the Kingston post office to hold for him to pick up on his way back to England had been removed. Early in February Dr. Strachan had gone to Cornwall on business connected with the settlement of the late Richard Cartwright's estate.[22] Along the way he conferred with some of his former pupils and their fathers about the man who in a few months

---

21*Ibid.*, p. 480. The preference on timber, doubled in 1810 during the war, was being revised.
22*The Banished Briton*, #22, p. 252.

had become the talk of the province. It was after this trip that some of the reports were withdrawn.

As Strachan was discussing with his friends his annoyance over the newcomer, Gourlay was listening to his land-developing relatives in Niagara. Their complaints about government land policy were turning to roars. The Honourable William Dickson was standing at his window daily, watching for buyers to come in from the United States in answer to his advertisements. January travel, if frigid, was fast and winter was the time when settlers often came from the United States to confer about a purchase.[23] Dickson's property in Dumfries was especially desirable for it was former Indian land with no clergy or crown reserves to retard its development. Now, as he listened in vain for the jingle of sleigh bells and the crunch of strange footsteps in the zero January snow announcing the arrival of prospective buyers at a dollar or two profit an acre, Dickson's face grew longer and longer. Reports of the change in government policy were keeping settlers away. Dickson was now angry. At a mess of the 70th regiment, of which he was commanding officer, he became tipsy enough to assert loudly that if things did not change he would rather live under the American than the British government. Thomas Clark was just as incensed. He told Gourlay that the conduct of the government was sufficient to justify rebellion. He said that Canada would not remain a British province for five years if management continued the same. All this made Gourlay resolve that in addition to forming a land agency he would set up a newspaper. He would locate it in the same centre as his land agency but would move to York during parliamentary sessions so that he could report assembly debates verbatim as William Cobbett was doing in England.[24]

It was now February 1818, and parliament was about to convene for the first time since its prorogation by Gore in May 1817. Fortified by the support of the opinions of his two legislative councillor relatives and all the township reports that were pouring in to him, Gourlay decided to write a second Address to the Resident Land Owners of Upper Canada. His sojourn in the province had given him ample opportunity to size up its political condition and he was ready to offer the people some advice, a thing he had not done in the first Address. The day after he finished writing it, he breakfasted on beefsteaks with magistrate Samuel Street, Thomas Clark's partner and a native of Connecticut.[25] He handed him the manuscript of his second Address. Street had not read far when he began to tremble. "What is the matter?" said Gourlay. "This must not be published," said Street, "I must take notice of this." Street said he

---

[23]*Statistical Account II*, p. 468 ff.
[24]*Ibid.*, pp. 470, 613.
[25]*Ibid.*, pp. 483 ff.

was bound by an Act of parliament to suppress such words, and calling
his wagon to the door, he whipped up his horse to go to consult with
Clark. As soon as Sam Street had driven off, Gourlay jumped on his horse,
and without asking any one else's advice he rode to St. David's where he
gave the first three paragraphs of his second Address to the *Spectator,*
saying that he would give in the rest the following week for publication.
Gourlay may not have possessed his own newspaper but the *Spectator*
was giving him as good service as if he owned it, and if he had had a paid
press agent, he could not have adopted a more effective device for stirring
public interest than by publishing only the first three paragraphs, as
follows:

TO

THE RESIDENT LAND OWNERS OF UPPER CANADA

Queenston, February, 1818.

GENTLEMEN:

I did myself the honour of addressing you through the medium
of the Upper Canada Gazette, of the 30th October last, and my
Address has been since widely circulated over the province by various
other channels. Its object was to gain the most authentic intelligence
concerning this country, for the information of our fellow-subjects
and government at home. The object was important: the means
employed were simple and fair: the effect it produced was palpable.

To lull the spirit of party, and quiet every breath which might
stir against a measure so chaste and efficient, I forebore all allusions
to political concerns. Conscious of being moved by the purest intentions, and desiring alike the welfare of this province and its parent
state, I dispatched a copy of my Address as soon as published, to be
presented to Lord Bathurst, and trusted, that by calm and dispassionate statement at home, the supreme Government would be best
persuaded to amend the errors of original institution. In these, I
conceive, lay the chief obstacles to the prosperity of the province; in
Canada I thought there was but one interest: in simplicity I said,
"Here we are free of influences."

Since then, three months have passed away. In this time I have
travelled more than a thousand miles over the province: I have
conversed with hundreds of the most respectable people: I have grave-
ly and deliberately considered what I have heard and seen: I have
changed my mind: and most unwillingly, must change my course
of proceeding. This country, I am convinced, cannot be saved from
ruin by temporizing measures, nor by the efforts and reasoning of any
individual. If it is to be saved, reason and fact must speedily be urged
before the throne of our Sovereign, by the united voice of a loyal

and determined people: if it is to be saved, your Parliament now assembled must be held up to its duty by the strength and spirit of its constituents; a new leaf must be turned over in public conduct; and the people of Upper Canada must assume a character, without which all Parliaments naturally dwindle into contempt, and become mere tools, if not the sport, of executive power."[26]

Mounting his horse again, Gourlay rode over to councillor Dickson's house to see what he thought of the second Address. Dickson roared his approval. Slapping Gourlay on the back he took him into a private room. "Mr. Gourlay," he said, "you must accept from me a deed of five hundred acres of land." Gourlay relished Dickson's approval but he was too wise to accept what might be construed as a bribe. Approval was reward enough. Let cautious Sam Street and Tom Clark get together! He and Willie Dickson were two of a kind. One evening, when the gentlemen were alone with their after-dinner wine, Dickson had boasted to Gourlay that he was the richest man in the province. When he talked of developing Dumfries, his hopes mounted sky high. Throwing back his head, he dashed a bumper of wine down his gullet and cried, "There goes another acre of land!"[27] How well Gourlay got along with forceful Dickson!

After this highly satisfactory interview, Gourlay again leaped on his horse. Not far from Dickson's house he was stopped by a messenger with a letter from councillor Clark begging him not to publish the Address he had shown to Street for fear that he and Dickson might be called its *fathers*. Gourlay had just come from Dickson and his jubilant offer of 500 acres of land. He had no fear for his child. Instead, he returned to Street's house to ask him to produce the Act that Street felt bade him suppress the second Address.

Immediately Street brought out the Upper Canada statute book and turned to the 44th Geo. III, the Seditious Alien Act that the legislature of Upper Canada had passed to keep the country free from republicanism. The war that had called it into being was over but Upper Canada had not repealed the Act. Gourlay took the book, read the Act, and then, like Sir Patrick Spens in the old Fife ballad, "A loud lauch lauched he". He could interpret the law better than His Majesty's magistrate! That Act, he told Street, applied only to aliens and he was a British subject, born in Fife. He was no alien. Like St. Paul he boasted, "I am a Roman and touch

---

26*Ibid.*, pp. 471 ff. For the ending "land of lice or liberty" see *The Banished Briton*, #18, p. 183, or *Letters to the Resident Land Owners*, p. 30. No copy of the *Niagara Spectator* for this date has been preserved.

27*Statistical Account II*, p. 493. Dickson's cavalier attitude to Upper Canada's land was exceeded on at least one other occasion. When Amherst Island near Kingston (granted to Sir John Johnson for loyalism) passed into Lady Bowes' hands, she gambled it away at cards. A. Lower, *Colony to Nation*, p. 167.

me if you dare."²⁸ Street read the Act again and was forced to confess that Gourlay was correct. The Act did apply only to aliens and he knew Gourlay was completely British. After that victory Gourlay proceeded serenely to the post office to see if there were any more reports waiting for him. He was keeping brother Thomas busy copying them so that he could leave a complete file behind him in case he were shipwrecked on the voyage home.²⁹

On the same day on which Colonel Smith had read the Speech from the Throne (that Gourlay intimated Chief Justice Powell had written for him), Gourlay wrote to his wife.

"Peter Hamilton, who arrived ten days ago, brought me the latest account of your welfare, and I turst you got all settled. . . . Mr. Clark and Mr. Dickson go off for York in a few days, to attend their duties in parliament. I follow them in a few days, and shall stay there two or three weeks to see how matters go on in the assembly, which has very important matters to discuss.

"I have a thousand things to say of men and manners, but it must be all left for conversation. Good God! let us meet, never again to be so long apart. I have a curious existence here; not unhappy indeed, but far from desirable. One thing, nobody can see my uneasiness; indeed, I am uniformly light-spirited to all appearance,—better, in fact, than I can account for. You must not be anxious though my letters do not reach you regularly after this time. Towards spring, the voyages are tedious—sometimes three months. I hope to God you will weather every difficulty till I get home. Sheep would be got in for the spring feeding, and 600 or 700 for summering. These Mr. Powell will provide. . . . God bless you, grand-mamma, and our dear children."³⁰

On the same day he wrote to London to his friend Major-general Sir Henry Torrens, a Fife gentleman who had been aide-de-camp to the Prince Regent. Maybe his relatives Dickson and Clark had been right after all when they told him that he would get no action from York but would have to apply to the Colonial Office for support for his immigration scheme. He told Torrens that under the prevailing land system hundreds of people soured and went to the States. He suggested that Sir Henry speak to the Colonial Office about his plan and asked him to request Lady Torrens to drop a line to her good friend, Mrs. Gourlay, to let her know that he was well and hoping to be home soon.³¹ Every

²⁸*Statistical Account II*, p. 490.
²⁹*The Banished Briton*, #33, p. 467.
³⁰*Ibid.*, #25, p. 313.
³¹*Statistical Account II*, p. 464.

day that passed heightened his determination to obtain redress for the refusal that York had given him. Maybe he would be the one who would be able to effect a change of policy towards Upper Canada. Robert Nichol had been in England for eight months now with documents detailing war and militia claims for land. Reports were filtering back about his continued failure to receive compensation for his war losses and a more realistic land policy. Nichol had failed so far but Talbot had been successful enough that he had treated former Lieutenant Governor Gore with marked rudeness.[32]

After he finished his letter to Sir Henry Torrens, Gourlay drafted a petition for his cousins to present to the assembly for him. He requested permission to appear before the Bar of the House to lay before it the findings of his western trip and his treatment by the land council.[33] His cousins thought the petition was a capital idea. They chuckled as they visualized the consternation that would appear on the faces of the members of the executive council, their superiors in government, when they read it. They even advised their cousin what suit to wear for his appearance at the Bar of the House. He was to wear his black suit, for the government was in mourning for the death of the twenty-two year old Princess Charlotte, only legitimate daughter of the Prince Regent and more beloved than the Regent's brothers who were now next in line of succession. Official mourning for a princess was black coat and purple breeches, but purple was a little hard to find in the colony. A black suit would do nicely when Mr. Gourlay made his appearance at the Bar of the assembly. He was to wait in Niagara until the two legislative councillors sent him word to come to York to witness the presentation of his petition.

On February 12 subscribers to the *Niagara Spectator* opened their papers to read the rest of Gourlay's second *Address to the Resident Land Owners of Upper Canada*. Some of it was conservative in tone, like the statement that the purpose of government was the protection of persons and property, and that it was not land hunger alone that caused Americans to come to Canada but dissatisfaction with their government. Most of it was critical. Canada, he wrote, could have even better government if the assembly watched its actions. The assembly should have impeached its departed lieutenant governor instead of expelling Durand for telling the truth in his address to his electors. Gore's private life was irreproachable, but his conduct of public affairs, particularly his prorogation of parliament "in a style unheard of since the days of Cromwell", was a standing jest. The king whom Gore represented could do no wrong but his ministers often erred. Gourlay said he had exposed the peccadilloes of certain

---

32There were imperial regulations that Gore disliked as much as Gourlay and Nichol did. He would exclude Americans, but he disliked the system of free grants of land. Q 317, p. 223, DA.
33*The Banished Briton*, #18, pp. 201, 219.

local ministers in order to make clear the need of Upper Canada for a
more liberal system of management. Certain revisions of the constitution
would ensure this. The constitution was man-made, and men could revise
it instead of regarding it as unchangeable. "Our constitution . . . is that
beautiful contrivance by which the people, when perfectly virtuous, shall
become all-powerful; but which reins back their freedom in proportion to
their vice and imbecility."[34] The present administration would not likely
embark on a policy of reform. "In my humble opinion, Gentlemen, there
ought to be an immediate parliamentary inquiry into the state of this
province, and a commission appointed to proceed to England with the
result of such inquiry. This measure should not be left to mere notion of
Parliament: it should be pressed by petitions from every quarter,—from
individuals and public bodies; . . . ".[35] He knew hundreds of farmers with
money in their hands, men who would keep the poor under their wings till
they could support themselves, men who would come to Canada instead
of going to the United States if the province would only sweep its house
clean of vermin and filth. He concluded: "It is quite natural for us to keep
our connection with home, but we cannot prefer a land of lice to a land of
liberty."

It was little wonder that the reading of this address had sent Street off
to Clark in such a flurry. James Durand had been expelled from the assembly less than a year ago for similar words. This was democracy rearing
its dreaded head. No demand for inquiry into government actions could
be tolerated. Furthermore, Street and his partner Clark were waiting
for the executive council to make a decision on a lawsuit over their claim
to the water rights of the entire Niagara Rapids. Robert Randall was dis-
puting their monopoly. If they could get a king's patent through the
lieutenant governor, their position would be strengthened and their
monopoly sustained.[36] Neither could afford to jeopardize their business
affairs in pursuit of a will-o'-the-wisp called Liberty.

Street's attitude to Gourlay's demand for a commission of inquiry was
the approved government attitude. In Britain Lord Castlereagh had even
tried to pass a Peace Alien Act in order to legalize government suppression
of criticism in peacetime as well as in war. Dickson was less cautious than
Street. Dickson sent back jubilant word from York that copies of the second
Address were in such demand that Mr. Gourlay was to bring with him all
the extra ones he could lay his hands on. Dickson could get twelve dollars
each for them in York.[37] Lice or not, the legislators were reading.

---

[34]*Statistical Account II*, p. 480.
[35]*Ibid.*, p. 474.
[36]*Ibid.*, p. 507; E. A. Cruikshank, "Postwar Discontent at Niagara in 1818", O.H.S.P.R.,
    1933, p. 25.
[37]*The Banished Briton*, #20, p. 219.

15

The Four Horsemen of Niagara

Gourlay had not concluded the first draft of his second "Address to the Resident Land Owners of Upper Canada", February 2, 1818, with "land of lice". He had ended it on the cheerful note that property would rise ten times in value if a liberal connection with Britain and a liberal system of home administration were established. He added that rather provocative conclusion when the last paragraph was being set, for he was wild with anxiety over a letter he had just received from home and furious at what he considered was the paltry patronage that was delaying him in Upper Canada.

On February 16 he had opened a letter from Jean dated November 16.

"Where are you, my dear Gourlay?—my heart has failed me at last, and I dread the worst that can befall me. I look upon my children, as if they were already fatherless, and pity myself as the most forlorn of widows. They tell me of lost letters—and letters detained—and of husbands who have been years without writing, and yet come home safe and kind at last; but nothing of all this seems to apply in my case;—for I am sure you will write as long as you are alive, and think me so—and so often, that it is not at all likely that all my letters can be lost or detained. What do I write for then?—Partly, because I can find no other employment in the least interesting, and partly, I suppose, because I still do hope, in spite of all my heaviness. . . . The people in the Parish wish very much for you back again, and a party of them were telling John, that they should set the bells a ringing, if they had you but home; . . ."[1]

He was also outraged by Administrator Smith's Speech from the Throne, a copy of which had just reached Niagara: ". . . I recommend it to

---

[1] *The Banished Briton*, #19, p. 214.

your consideration how far it may be expedient to assist the immigrants by providing the means to defray the expense of the location and grant of land bestowed upon them by His Royal Highness, the Prince Regent, in His Majesty's name." There was no mention in the Speech from the Throne about absentee owners, compensation for war losses, St. Lawrence canals, lack of capital for development, nothing that would answer the known demands of the province. In the *Niagara Spectator* of February 17 Gourlay published all his correspondence with York officials regarding his own application for land.[2] He knew, he wrote, about how far the government was prepared to go to help immigrants! He had migrated from the United Kingdom; had the government countenanced him? He had proposed a better scheme than Smith's to encourage immigrants, one that would put money into the treasury instead of taking it out. He was still waiting for an *honest* answer to his petition. What was wrong with York? "Has it devoured too much land and gone to sleep while the crude mass is yet undigested? . . . How is it, that it has duties to perform and cannot perform them?" He considered the conduct of public affairs at little York dull, dirty and disgusting. Gone was the reasonable tone of the two Addresses. This was war, and he would out-Cobbett Cobbett.

He abandoned generalities. He told Dr. Strachan, who had advertised in the *Upper Canada Gazette* that he would give a course of lectures on natural philosophy for the sum of two guineas and use the proceeds to paint the church, that he would lecture on political economy, and not only paint the church outside but "clear it of cobwebs within: perhaps, I might fit it with a steeple and a bell, and make it look decent beside the *palace* of its pastor".

He told Administrator Smith that the first act of the legislature that was so fond of saying it had no money should be to tax wild land, or borrow money from Lloyd's of London. In view of the wonderful prospects it possessed, why did Upper Canada hesitate to bring in capital? Gourlay knew that Britain was heavy with capital to invest. "Capital," he said, aware that there was nothing mysterious about it, "is much a thing of idea and rests on confidence. The British funds are nothing but ideal property, held by confidence in the future proceeds of skill and industry." If a man like Law could raise money to find gold on the banks of the Mississippi, and another (though Gourlay did not cite this prime example of British readiness to invest before investigation) to sell skates in Rio de Janiero, surely Upper Canada could find capital for sound commercial projects like roads and canals. Administrator Smith had suggested a vote of money to help immigrants pay fees to the land office whereas

[2]*Ibid.*, p. 211 ff.

"their great want was ready service on the part of the Land Council, and land in situations where it was possible to clear it".[3]

Then, in his anger, he took part of his wife's letter (the one quoted above) and printed her loving words for all eyes from Sandwich to Quebec to see. He did it because "In a tumult of feeling,—in a paroxysm, I thought of my wife's anxiety,—I thought of hundreds of emigrants who had been vexed and disappointed, and torn to pieces, by the vile, loathesome and lazy vermin of Little York. I attacked the whole swarm, and flung into my letter an extract from my beloved and distressed wife. The bolt was discharged, and kindled inextinguishable flame. Little York! Little York on fire!"[4]

When this issue of the *Spectator* reached York, Clark was so aghast that he refused to present Gourlay's petition asking for an inquiry into the affairs of the province. He, Dickson, Beasley, Captain Stuart and many others had had to wait for their personal development schemes to materialize. Why should not Gourlay? Though Gourlay's anger was directed against slack administration, Clark considered Gourlay's letter a breach of good manners. "Every man of respectability will be shy of you," he wrote from the capital and suggested that Gourlay stop embarrassing his cousins and go home to Britain at once, not by way of York, but by way of Sacket's Harbour on the American side of Lake Ontario.[5] When did Gourlay ever play the coward? "Go by way of Sacket's Harbour," he exploded again in the *Spectator*, ". . . My dear friend Clark—the bravest man in Upper Canada . . . bids me go by way of Sacket's Harbour! . . . I'll blow Little York in the air, and 'every man of respectability' in it, before I go by way of Sacket's Harbour!" He did not have to rely on Clark to present his petition to a colonial legislature. He bundled it up with all the letters he had published in the papers so far and sent them off to Lord Bathurst.

Then he continued, picturing what had happened in York to make Clark and Dickson change their minds about him.

"You dined on Monday with the Honorable This, and the Honorable That, Doct. Thingam, and others. They were no sooner met, than the *Spectator* was talked of.—'Well, what do you think of it now?' says the Doctor: 'didn't I tell you from the beginning what this fellow was? Here is a private correspondence published to the world:— what

---

[3] *Statistical Account II*, p. 565.
[4] *Ibid.*, p. 214; Twentieth Century Canada has castigated Gourlay for publishing these extracts from his wife's letters. In 1813 the Prince Regent published his correspondence with the king who refused to let him lead the army in person, to prove that he was no coward. A few months before this, letters from the Princess Charlotte to her mother, the Princess Regent, were published in connection with the political game of Tory support the king, Whig support the queen. This was the great epistolary age.
[5] *The Banished Briton*, #20, p. 217.

gentleman would do that? . . . Why did you lodge and feed him? Why did you not let him die of musqueto bites?' 'Why did you save him, thus to laugh, even at the newspaper musquetoes of Canada; and to silence the *majoring* of half-pay officers?' . . . Do you remember my declaring, before I visited York, that I would have no introduction to this little man? . . . Let him apologize to Lord Selkirk: let him never again intermeddle with the free commercial speculations of any individual: let him never again blow up strife among mankind: let him give up dabbling in politics, and trifling with philosophy: let him resign his seat in council, where no priest should ever have a place: let him get into a penitentiary: let him stick to his altar, where his utmost zeal and ability has scope; and then we shall leave him to God and his conscience. . . . I expose nobody without good cause. . . . Let every thing, and every man, be *exposed* for the sake of truth, justice, and the law. . . . The salvation of this province depends on a few of you, now at York, sacrificing every little consideration—tearing yourselves from the fascinations of personal feelings, and *really* doing your duty."[6]

Robert Gourlay was now a full-blown gadfly. He was asking the administrators of Upper Canada to sacrifice for the good of their country in peacetime as they had in wartime, the type of request that usually goes unheeded.

The *Spectator*'s columns soon contained more than advertisements for lost horses and remedies for itching. Major Richard Leonard wrote to say that he was quite content with affairs as they were.[7] Gourlay answered that he should be, for he was retired on half pay on a choice farm in the well-settled Niagara peninsula. A "Resident Landholder" wrote that he was happy with his dear, his native land, and did not need a "counterfeit saviour" like Mr. Gourlay. The *Traveller*, who wrote a weekly column of innocuous observations on differences between the new country and the old, commended Gourlay for trying to wake up a province that till now had shown no interest in political discussion, but he said that it was useless to expect "the ruffian remnant of a disbanded regiment, or the outlawed refuse of some European nation" to have anything intelligent on their minds.[8] Gourlay filled half a newspaper page defending his efforts and the common people of Upper Canada in the face of the *Traveller*'s defeatism:

> "The times are out of joint! O cursed be I
> If, born to set them right, I never try!"

---

[6]*Ibid.*, #20, p. 219 ff.

[7]*Ibid.*, #18, p. 191. Gourlay claimed that Leonard's wife wrote the letter. Soon after, Leonard was made sheriff.

[8]*Ibid.*, #22, p. 262.

This letter turned the *Traveller* (Dr. Howison) against Gourlay. He threatened to rouse the Niagara magistrates to issue a warrant for Gourlay's apprehension. After they had committed him to jail, wrote the *Traveller*, "they will perhaps be able to reply to his 31st query, by stating, whether or not his imprisonment 'retards the improvement of the Province, and if his enlargement would contribute to the same.' "[9]

Gourlay was contemptuous of such threats for he had read many worse diatribes in the London papers. Colonel Smith of Upper Canada may have used the utmost delicacy, for instance, in the Speech from the Throne when he referred to "the melancholy death of Her Royal Highness, Princess Charlotte Augusta, in circumstances of peculiar interest to the nation," but the London satirist, Peter Pindar, seized more lustily on the situation:

> "Yoicks! the royal sport's begun!
> I' faith and it is glorious fun
> For hot and hard each royal pair
> Are at it searching for the heir!"[10]

Gourlay was rather enjoying himself too as he sought to relieve the poverty of the British poor by pricking the people of Upper Canada to action. He was furious about existing conditions but this was also a game of wits. He wrote Clark not to be hurt at what he was saying. "Enquire into the practices of our first statesmen at home, and you will find that *as public men* they abuse each other like pick-pockets; and, the very next day, crack a bottle together, in the utmost conviviality of private friendship. I published a letter of the President's marked *private* and for what reason? It contained not a word of consequence to conceal. The word *private* was written upon it merely to distinguish it from an official communication."[11]

York was not London where Gourlay had learned his craft. Its leading men were not nobles who could battle like fishmongers among themselves and still retain their titles, or royal dukes who could holler like huntsmen in the House of Lords and still be the first family of the nation. Most of the people of York were small town people on the way up, uncertain of their table manners, unable to change the rough accents of their origin like Dr. Strachan, or trading on crumbs of patronage dropped down the social scale. There was no comfortable obscurity here. No one walked

---

[9] *Ibid.*, p. 246.

[10] George III had thirteen sons and daughters. Princess Charlotte, only child of the Prince Regent and Caroline of Brunswick, married to Leopold of Belgium, was beloved of the nation. Her death in childbirth was a blow, for legitimate royal offspring were rare. To meet the situation, a clutch of middle-aged dukes was ordered to leave their current loves and marry.

[11] *The Banished Briton*, #22, p. 219.

along a street in York without sensing an imperceptible movement behind the lace curtains. The chief sport in York and in the whole province was gossip. Who of these people would crack a bottle in private with the man who trounced them in public? This was a community in which speech had to be guarded for enmities were made easily. The Jarvises and Ridouts in York were a prime example. Some one spread a rumour that the Jarvises had not paid their daughter's school fees in Quebec. Then the Jarvis carriage splashed the Ridouts on their way to St. James church. The sons fought a duel (July 12, 1817) in which young Ridout was killed.

In York the difference between public and private life was narrow, not wide as in the sophisticated environs of London. There, where his campaign had been directed against an impervious nobility, a centuries old parliament, and a country accustomed to political spats since the days of Magna Carta, the tone of his pamphlets was less that of virulence than bewilderment that an irrational administration should be allowed to perpetuate itself. Here, where the government was in the hands of a colonel, a schoolmaster hired recently by his wife's uncle, and his wife's cousins, he had become overnight a master mudslinger, and he was resented as such.

Though Thomas Clark urged Gourlay to fade away unobtrusively, the people from the Western, Gore, London or Niagara Districts were not asking to have their reports returned to them, and in Kingston "Junius" was writing to the *Gazette* unmasking the man who had first expressed his disapproval of Gourlay's proposed statistical account. "His attempts, however, appear to have been thwarted, particularly by the jealousy and prejudices of one man, whose illiberality is becoming proverbial, and who, unfortunately for the country, seems to have it in his power, as well as his disposition, to counteract the most useful undertaking, if not originated by himself. . . . There are times when it is the duty of men of abilities and independent circumstances to check malpractices by a free and candid examination of them to the public."[12] The Montreal *Herald* (Feb. 28, 1818) agreed: "Mr. Gourlay was told, that being a stranger . . . it was indecorous in him to meddle with public affairs in the face of so many superiors; but the truth seems to be, he was found possessed of superior talents, and that was enough to get him enemies."

On February 23, 1818, Administrator Smith was not disposed to be as dispassionate as Junius and the *Gazette*. He sent Lord Bathurst the pages of the *Niagara Spectator* that contained Gourlay's second Address, a résumé of Simcoe's Proclamation, Bathurst's own statement in 1815 disapproving American immigration, a copy of the Seditious Alien Act of 1804, and a diatribe against land speculators who had bought Indian land

12*Kingston Gazette*, March 17, 1818.

and whose plans for development had been discommoded by imperial prohibition. "They have now found support out of the House in a reformer from the United Kingdom, whose declarations in the Provincial *Gazette* are not the less inflammatory amongst an ignorant population from the want of Truth, reason and decorum.

"This man, Robert Gourlay, one of whose lucubrations I have the honour to enclose, has no property or fixed residence in the colony and is only known as a relation to Mr. Dickson, proprietor of a Grand River Township."[13]

All this he laid before Lord Bathurst in his wobbly, illegible writing, revealing that the main reason for his opposition to Gourlay was hatred of the swift methods of land developers.

When Smith mailed this plea for help to Bathurst, he was nearing the end of his ability to deal with the parliament to which he had read the Speech from the Throne that had sparked Gourlay's second *Address to the Resident Land Owners of Upper Canada.* Instead of being cowed by the prorogation of its first session by Gore, the assembly that Smith had summoned for its second annual session on February 5, 1818, was preparing to renew the battle for the control of the people's purse.

On February 6 the assembly re-introduced the bill "to assess land of absentees . . . and for rendering them liable to do Statute Labour on the highways". It introduced a bill to repeal the Act passed in 1816 appointing and paying the salary for five years for a provincial agent in London. The legislative council returned it with the amendment that the agent's term be extended for life.[14]

The assembly wished to pay Jonas Jones the £100 he had been promised for his expenses in accompanying the commissioners who had negotiated the division of tariffs collected at Côteau du Lac. The legislative council adopted a method of dealing with bills that it disapproved that would soon become a habit: it let the bill "lie asleep", hoping that the Commons would grow tired.[15]

The assembly showed its teeth. The civil list estimate of £2,500 in 1816 and £8,251 in 1817 had increased this year to £12,603. When the assembly was asked to vote this amount, it refused unless the money voted for the two preceding years was accounted for. It went further. It introduced the Civil List Repeal Bill to rescind the £2,500 voted in 1816. When

13Smith to Bathurst, York. Feb. 23, 1818, CO 42/361, pp. 23-25.

14*Jls. Leg. Ass.*, Ontario Bureau of Archives, Ninth Report, Feb. 6, 1818, p. 433.

15*Ibid.*, Feb. 16, 1818, p. 446. They justified their action on the grounds that the preamble contained the falsehood that Jones had already been paid. As a collective body they pretended ignorance of the truth, but as individuals they knew Jones had not been paid. This recalls Gourlay's expulsion from the Bath Society for saying that they "were individually good fellows but collectively, great rogues". Strachan's phrase "lie asleep" is found in his *Letter Book*, p. 157.

Administrator Smith reported that in addition to the £800 granted to Scott as a pension, £400 had been paid to Chief Justice Powell as his salary as Speaker of the legislative council, the assembly passed the Civil List Repeal Bill, 9-8.[16]

The assembly was not yet finished with money matters. On March 13 it asked for an account of the spending of £2,000 worth of duties and militia fines collected in 1817. It passed a bill authorizing the sale of deserters' estates, the proceeds to be applied to the settlement of war losses. It appointed a committee to confer with the legislative council about drafting a joint petition to the Prince Regent on the subject of land grants to the Volunteers and militia.[17]

The assembly sent to the legislative council a bill regulating trade with the United States. The upper house returned it with amendments claiming that the lower house had misinterpreted certain regulations. Led by the reinstated James Durand, the assembly sent the bill back again, claiming that the upper house could not tamper with a money bill. The legislative council replied that the origin of supplies was indifferent. After much altercation, the legislative council declared that it would not amend money bills but merely reject them, a procedure that would let the electorate know who was responsible for failure of legislation. The assembly was not deceived by this apparent concession, for it knew that the issue of unauthorized expenditure of funds was still unsolved.[18]

Executive councillor Dr. Strachan was not uninterested in this interchange of blows. He considered that the legislative council's power of rejection of money bills would give it sufficient control over the assembly without quibbling over the right to amendment. He was not advising the council to yield any of its power, for he still wanted a seat on it. Neither was he championing the supremacy of the assembly, for he was writing chattily to the secretary of the governor-in-chief in Quebec that the cause of the new crisis was the "revival of that old dispute about interfering in money bills".[19] How bothersome could a backwoods assembly be? was his implication.

The backwoods assembly knew exactly what it was doing. It was determined to have the exclusive right of initiating money bills. It was

---

[16]*Ibid.*, Feb. 16, pp. 450, 452.

[17]*Ibid.*, pp. 453, 527, 434, 523.

[18]*Ibid.*, pp. 542, 546. In February, 1812, Richard Cartwright had chided the assembly for affecting to copy "the illustrious House of Commons with which they arrogate an Equality". (Cartwright Letter Books, p. 354). The background for this dispute may be found in B. Kemp, *King and Commons*, 1660-1832 (MacMillan, 1957) pp. 71 ff. In the 1670's the Commons claimed that money bills could not be amended in the Lords. Under William of Orange, parliament began regularly to state for what purpose money was granted.

[19]Strachan to Gore, York, April 10, 1818, G. Spragge, ed., *The John Strachan Letter Book*, p. 157; to Halton, p. 159.

engaged in the ancient struggle for the power of the purse, the only peaceful weapon that democracy has devised against autocracy. On March 23 it passed a resolution that "as all charges or burthens whatsoever upon the people of right begin with the Commons, so, they cannot be altered or changed by the Honourable the Legislative Council. . . ." The next day this answer was flung at them by the legislative council: "the Legislative Council and House of Assembly of the Province of Upper Canada are co-ordinate branches of a Limited Legislature, constituted by the Statute enacted in the thirty-first year of His Majesty's reign. . . ." The assembly countered by quoting Lieutenant Governor Simcoe's happy claim at the opening of the first session of Upper Canada's first parliament, September 1792, that Canada now enjoyed the "image and transcript" of the British constitution. The assembly reminded the "nobles" snidely that that part of their minutes was missing, destroyed by the enemy in the late war.

The assembly resolved itself into committee of the whole to plan its next move. Led by James Durand and Richard Hatt, both of whom had written critical answers to Gourlay's thirty-first question, the assembly asked Administrator Smith to approve supplies by address, without the concurrence of the legislative council. This Smith refused to do till he had consulted His Majesty's pleasure thereon.

The assembly refused to wait half a year for an answer. That same day it adopted two petitions. The first was addressed to the Prince Regent, ecclosing an abstract of proceedings of the present legislature and remarking on "the evil that must necessarily result from the Legislative and Executive functions being materially vested in the same persons". It asked the Prince Regent (humbly) to remind the legislative council that it could not usurp the rights of His Royal Highness' faithful Commons. The second petition was directed to Administrator Smith begging him to transmit their petition to the Prince Regent.[20] As the assemblymen had a fair idea of what the administrator's reaction would be, they had no sooner executed this courageous measure than they lost their nerve and most of them skedaddled out of York. They figured that Colonel Smith could not issue a warrant to arrest so many for contempt of parliament, nor could he prorogue them if they were not there. Their courage was considerable, but it was not sufficient for them to declare themselves a Long Parliament or to take a Tennis Court Oath not to disperse until their status was clarified.

They were right in guessing that Administrator Smith would not receive their petitions amiably. He told his councils that he would submit no petition to the Prince Regent that had not first been approved by him. Though he had a reputation for delay that outclassed Fabius Cunctator, he belied his reputation this time, for Gore had set him an example

---

20 *Jls. Leg. Ass.*, March 30, pp. 562-4.

the year before. He prepared a short speech that ended with "Finding no probability of any concert between the two Houses, I come reluctantly to close the session with its business unfinished." On April 1, 1818, after he assented to certain laws, he delivered it to the rump of the assembly that appeared before him when summoned. It was an illegal prorogation for there was not a quorum present, but how else could he dismiss an assembly that had already "flown the coop"?

The people of Niagara were well aware of the war that was being waged between the commons and the peers at York. Therefore when they heard the distant sound of cannon shots coming from across the lake, they did not attribute it to an American invasion. They knew that parliament had ceased to sit, and they were agog to know who had won the struggle. On April 2 a crowd was at the dock to meet the parliamentarians who were in the indeterminate position of returning neither with their shields nor on them. Wild horses could not have kept Gourlay away, and when he learned that the members were returning because parliament had been prorogued before the impasse was resolved, he scoffed, "The fools were dismissed on their own proper day," and made his way towards Dickson who was still arguing heatedly with Mr. M'Donell, Smith's brother-in-law. Dickson gave Gourlay a copy of the speech with which Smith had closed the session and then delivered his opinion of the matter. Dickson was angry with his superior, executive councillor Strachan, who had opposed the stand of the legislative councillors, and with his mouth and face screwed up to mimic the dominie's contorted face, he imitated his harsh Aberdeen brogue: "They're wraung! they're wraung!"[21] In an effort to solve the deadlock Dickson had submitted resolutions to the legislative council but Chief Justice Powell had had his accepted instead. Dickson vented his anger against the legislative assembly too, for he was firm in his belief that the legislative council should be able to amend its bills. He pretended, however, that the upper house had tried to change nothing but inconsequentialities: "What! shall we allow bills to come up to us with bad spelling and bad grammar, and not alter them? No: it would never do to disgrace the statute books in this way."[22] Dickson gave his rejected resolutions to Gourlay to take them to the *Spectator* for publication, provided, he said, that Gourlay refrained from making any comments on them.

Gourlay did not bother with Dickson's resolutions, for he had his own opinions to publish. This was a crisis of vast importance. He left Dickson's house, returned to his lodgings, and at one sitting wrote a third *Address to the Resident Landowners of Upper Canada*. On April 9, 1818, it filled half a page of the *Niagara Spectator*.

---

[21]*Statistical Account II*, p. 574.
[22]*Ibid.*, p. 567.

"Gentlemen,

Your Parliament is broken up!—and a second time broken up, from employment of the most vital import to the honor and well-being of the province!!—Good God! What is to be the end of all this?

"For my own part, Gentlemen, I had little hope of satisfaction from the sitting of Parliament, after perusing the Administrator's speech from the throne; and this little was entirely extinguished with the disgusting reply made to that speech by your Representatives. . . . Did we, the offspring of early civilization—the first hope of genuine liberty . . . come to this new world, only to witness the degradation of our kind . . . Surely, British blood, when it has ebbed to its lowest mark, will learn to flow again, and yet, sustain, on its rising tide, that generous—that noble—that manly spirit, which first called forth applause from the admiring world. . . . It has been my fate to rest here nearly two months, viewing at a distance the scene of folly and confusion. . . .

"Gentlemen, the constitution of this province is in danger, and all the blessings of social compact are running to waste. For three years the laws have been thwarted, and set aside by executive power;—. . . your Commons and your Peers have quarrelled, and the latter would assert that the constitutional charter of Canada may be trifled with. What is to be done? Do you expect anything from a new Governor? . . . Do you expect anything from a new set of Representatives? It is not the men, it is the *System* which blasts every hope of good; and, till the system is overturned, it is vain to expect any thing of value from change of, Representatives, or Governors.

"It has been the cant of time immemorial to make mystery of the art of government. The folly of the million and the cunning of the few in power, have equally strengthened the reigning belief; but, it is false, deceitful, and ruinous. The people of every nation may at any time put down, either domestic tyranny or abuse,—they may, at any time, lay a simple foundation for public prosperity: they have only to be honest, and, in their honesty, bold.

"In my last address to you, I said that the British constitution was "that beautiful contrivance by which the people, when perfectly virtuous, shall become all-powerful." . . . The British constitution has provided for its own improvement, in peace and quietness; it has given us the right of petitioning the Prince of Parliament; and, this right, exercised in *a proper manner*, is competent to satisfy every virtuous desire. . . ."[23]

Since the king's representative would not transmit the assembly's

---

[23]*Ibid.*, pp. 581-587.

petition to the Prince Regent, the people must petition the Prince Regent directly. They must meet in every township. Each township would elect a representative and clerk. The township representatives would assemble in each District on an appointed day to draft a petition to the Prince Regent. Each district meeting would choose a representative to meet in a provincial convention in York that would revise and approve the draft petition and choose commissioners to carry the final petition to England. Every man who endorsed the plan would give a dollar towards the expense entailed in drafting and conveying the petition to London. This convention would do what parliament would not. Though "the rights of parliament could be trifled with, those of the people of Upper Canada are not so easily set at defiance."[24] This method, he wrote, would lay the foundation for Upper Canada to become the most flourishing and secure spot on the habitable globe. Righteousness would exalt the nation. He set Monday, April 13, at Mr. Rogers' coffee house, as the date when the township of Niagara would hold the first meeting.

Before Gourlay took this third Address to the printer's, he showed it to Dickson. Dickson radiated good humour over it. He joked that he would send his black servant to the first Niagara meeting. "I returned the joke, by assuring him that the black servant with a dollar in his hand, should be considered as good a man as himself."[25] That week Dickson held a dinner at his house to discuss the Third Address. Before the evening ended, he and Clark led the others in subscribing their dollars towards the expenses of sending a petition to England. Gourlay had accomplished a good deal since the assembly had been prorogued on April 1.

On April 13 when he was walking to the coffee house for the meeting of Niagara Township, Gourlay was accosted by a member of parliament who objected to the criticism in the Third Address of the ineptitude of the parliament of which he was a member. He called Gourlay a traitor and leader of seditious meetings, but Gourlay refused to be provoked into a temper and continued peacefully to the coffee house.[26] David Secord, M.L.A., was appointed chairman of the meeting. The Third Address was read and approved. Robert Hamilton, Junior, was elected representative to go to the Niagara District meeting. Monday week was set for the meetings of the other townships of the Niagara District, and Monday fortnight for the District meeting at St. Catharine's. Four days after the publication of Gourlay's Address the organization of this district was complete.

Now councillor Clark began to have second thoughts. He talked with others like the member of parliament who had accused Gourlay of sedition. He, like Street at the thought of the Seditious Alien Act, began to tremble

24*Statistical Account II*, p. 587.
25*Ibid.*, p. 575.
26*The Banished Briton*, #22, p. 261.

nervously and to fear that his dollar might commit him to endorse whatever the Niagara District meeting might decide. He read again with apprehension that part of the Third Address that said, "till the system is
overturned, it is vain to expect any thing of value from change of, Representatives, or Governors". He then denied that he had given Gourlay a
dollar for his meetings, and on April 18 he distributed a handbill declaring
that the only project of Gourlay's that he had endorsed was the publication
of a statistical account of the province. He quoted from the common law of
the British empire, saying that the forthcoming meetings might be seditious
like the ones that had been forbidden by the Act of 1793 under which both
Irishmen and Scotchmen had been deported to Australia. He warned everyone "to weigh well how they attend to visionary enthusiasts" like Mr.
Gourlay. Then he called a meeting at Rorbach's tavern, Stamford, April
20, and read an address that he had written in his tidy, precise handwriting
spiced here and there with curlicues. He warned people to look well to
their dollars, for Gourlay had come to Canada to borrow money from
him and might use their dollars to pay his debts. "He talks of being the
Saviour of the country: to me he savours more of Beelzebub."[27] Then,
because Gourlay publicized a letter signed by W. G. Hepburne and certified
by Thomas Dickson and James Kerby that Clark had given him a dollar,
Clark tried to wiggle out of his denial, saying that he had given the dollar
out of pity for Gourlay for he thought it would be the only dollar he would
get.[28]

Neither Clark's handbill nor his Address could stem the rush to Gourlay's meetings. Five were held on April 20. The Grantham meeting read
both Gourlay's Address and Clark's. It preferred Gourlay's. It denied that
its purpose was seditious, saying that the 1793 Act applied to a time when
open rebellion was taking place in wartime. They said they only wished
to petition the mother country peacefully and quietly. They said they were
all loyal men. They elected William Hamilton Merritt as their representative and George Ball their clerk and put William Chisholm and Amos
M'Kenney (editor of the *Niagara Spectator*) on the continuing committee.
At the conclusion when Gourlay came in, they all ordered dinner and
drank copious toasts to their cause. Clark and his fears seemed to have little
support in Niagara, and Gourlay and his optimism a good deal.

The opponents of these meetings spread a rumour that ordinary people
could not petition the Prince Regent, only parliament. George Hamilton
then published a notice saying that he was asking a lawyer's opinion on the
right of the people to petition. Gourlay issued "An Address to the Worthy
Inhabitants of Niagara" declaring that their right to petition was un-

---

27April 20, 1818, Upper Canada Sundries, vol. 32, p. 142.
28*Statistical Account II*, p. 590.

questionable. He himself, he wrote, had been in the House of Commons when Lord Folkestone had presented a petition from one of the Spa Fields meetings. Clark was just trying to frighten people out of their rights. Clark was accusing him of sedition. "If I am guilty of sedition, why does not the Hon. Thomas Clark do his duty and bring me to trial?"[29] Clark could not, for he had done nothing seditious. He knew what sedition was. He had been at the Edinburgh trial of Thomas Muir, one of the chief offenders under the Act of 1793, but Muir had been charged with having warlike arms and holding secret meetings.[30] None of the township meetings was secret and no one was stockpiling arms. He contradicted Clark's claim that the constitution of Upper Canada could not be changed. Gourlay said that a parliament had made it and a parliament could change it.

With all this assurance from Hamilton and Gourlay, eight more meetings were held. At the Grimsby meeting Jonathan Wolverton was in the chair and Dennis Wolverton, the clerk, could not refrain from noting in the minutes the smiles of approbation and complete harmony of the meeting.[31] On May 1818, fifteen representatives met in St. Catharines to consider a draft petition. It asked for speedy compensation for war losses, award of promised land to the militia, amendment of the system of favouritism and patronage in disposing of crown lands, removal of fear of arbitrary power, and the appointment of a commission composed of men above base influence to inquire into the reasons why the province pined in comparative decay.[32] A committee of four was appointed to polish it, print it, and lay it before the public for one month for further suggestions. The *Niagara Spectator* published it on May 14, impressively double columned, and accompanied by an editorial speculating on the feelings of His Royal Highness when he learned that arbitrary power had grown to such a height in one of his remote dominions "that even magistrates crouched beneath its dismal shade".

The Niagara committee printed 1,000 copies of a pamphlet called *Principles and Proceedings of the Inhabitants of the District of Niagara for addressing his Royal Highness the Prince Regent*.[33] It contained the text of the petition, minutes of meetings held, all relevant correspondence printed in the *Spectator*, and Gourlay's third Address. Succeeding issues of the *Spectator* were filled with column after column of the names of 900 people who had subscribed their dollars to the cause of sending to the Prince Regent the petition asking for an inquiry into the affairs of Upper

[29]*Ibid.*, p. 599.
[30]*Ibid.*, p. 600; *Chronicles of Canada*, p. 6.
[31]*Statistical Account II*, p. 603.
[32]*Statistical Account I*, pp. 571 ff.
[33]This is said to be the first book printed in Upper Canada.

Canada. One issue alone listed 300 names. In one month Robert Gourlay had done this in the Niagara District with a population of 12,548.

Some said this response came only from peevish grievance-mongers. Some support did come from petty complainers who wanted something small like adjustment of a property line or road duty, but were there 900 chronic malcontents in the Niagara District? These men knew that the assembly was being prevented by arbitrary dissolution from doing its duty. They knew that the major work of the legislative council in the last session, in addition to rejecting the bills of the assembly, had consisted in composing a loyal address to the Prince Regent telling him how well everything was going in his colony of Upper Canada. They knew it was four years since the war had ended and important affairs were still not settled. If no settlement is made, people are left helpless like beetles on their backs. If the local government would not serve them, they would look elsewhere. Talbot and Nichol had gone to London for their own selfish interests.[34] Under Gourlay's leadership, the next men to go to London would be representatives elected to present the wishes of a large section of the province.

Early in May four horsemen paid by the Niagara committee rode out of the Niagara District to distribute *Principles and Proceedings of the Inhabitants of the District of Niagara for addressing his Royal Highness the Prince Regent* to four other districts of Upper Canada.[35] Gourlay was entrusted with the distribution in the four eastern districts, Midland, Johnstown, Eastern and Ottawa, for he had already planned to go there to collect his township reports from the post office in Kingston. For the first time in the existence of Upper Canada, every individual would be asked to give his opinion of the government of the province.

---

[34]In London Nichol suggested that he might receive a pension from the permanent grant of £2,500 which he said was proposed by him. Finally a pension of £200 per annum was granted him as compensation for war losses and services. As the Treasury commissioners suggested that he be paid out of the forfeited estates fund, it was a doubtful triumph.
[35]*Statistical Account II*, p. 575; *The Banished Briton*, #36, p. 334.

16

Hitching-Post Campaign

On May 18, 1818, Robert Gourlay arrived at Kingston in the Midland District accompanied by young Peter Hamilton who had recently returned from school in Scotland under the care of his aunt, Mrs. Henderson. They had a pleasant trip across the lake in the company of Chief Justice Powell who, with no perceptible anti-American rancour, was on his way to Boston to settle a family estate. The affable Mr. Powell could not possibly have been unaware of Gourlay's reason for going to Kingston. Strachan had been glad to see Powell leave York, for now he would have no interference in influencing Administrator Smith to take action against the meddlesome newcomer.[1]

As soon as he had settled into Walker's Hotel in Kingston, Gourlay sent young Hamilton to the post office to collect the township reports. He returned with the news that postmaster John Macauley had allowed lawyer Daniel Hagerman to remove from the post office the Adolphustown report for which Hagerman had acted as secretary. This had happened immediately after Strachan's February visit to Cornwall. Gourlay was furious. He knew that British postal regulations declared that all mail became the property of the person to whom it was addressed the instant it was deposited in the post office.[2] Why else would he have ordered his mail to be held in Kingston so that he could pick it up on his way home? He wrote a letter to the *Kingston Gazette* protesting Macauley's negligence as postmaster.

Gourlay brought himself up-to-date on another matter as well. Meetings had been held in Charlottenburgh, Lansdowne, Ernesttown, Sophiasburg, Adolphustown, and Prescott on unrecorded dates, likely in January, the time of the regular township meetings. He now had reports from

---

[1] Strachan to Col. Harvey, York, 22 June, 1818, G. Spragge, ed., *The John Strachan Letter Book*, p. 163.
[2] *The Banished Briton*, #13, p. 116.

the first four but not the last two. Prescott was the centre of Augusta
Township, the area settled by the first loyalists who had been evacuated by
ship from New York at the close of the American Revolution and given
choice St. Lawrence land. The annual Augusta township meeting had appointed a committee of eleven people to draw up its report. Among them
were Solomon Jones, district court judge, John Bethune, Strachan's successor as Anglican minister at Cornwall, and John Simpson, a recent arrival
from England. After Strachan's February visit the Augusta committee
met, but not to compile a report. In the March 3 *Kingston Gazette*, the
committee published a statement declaring that the First Address contained "principles inimical to the peace and quiet which the Inhabitants
of this Province happily enjoy".[3] The report from Charlottenburgh was
held back by one local member of parliament until another member insisted in April that it be delivered. When news of this opposition in the
eastern townships reached him, Gourlay wrote to the *Kingston Gazette*
on April 28 asking why he had not yet received reports from townships like
Augusta. Maybe, he needled, it was the unlettered condition of the eleven
yeomen who had been asked to write it.

The eleven yeomen, the majority of whom were former pupils of Dr.
Strachan, immediately enlisted the talents of the most learned among
them, John Simpson, who had not been educated by Dr. Strachan. Simpson answered Gourlay in the columns of the *Gazette*: "From the mis-
anthropic picture you have drawn of the British Government, you remind
me of Hegesias, the philosopher of Cyrene, who drew such a melancholy
picture of human life that many of his audience *slew* themselves in despair.
. . . I do not apprehend the like danger from your lucubration, nor can I
congratulate you *upon a single instance of suicide in the committee, or
even a visible appearance of dejection.*" Simpson took exception to Gourlay's criticism that Britain "annually sends abroad 20,000 to be slain". He
would not allow any criticism of his mother country:

> Poor England! thou art a devoted dear,
> Beset with every ill but that of fear, . . .
> Undaunted still though wearie and perplexed
> Once Chatham sav'd thee, but who saves thee next.

Gourlay wished to open the country freely to immigrants, but Simpson
feared that such a liberal policy "might make a Spa-Fields in Canada . . .
and provide you an auxiliary in a Cobbett, a Hunt or a Watson". Simpson's letter was published just two weeks before Gourlay's arrival in
Kingston.[4]

Ignoring Simpson's letter, Gourlay proceeded to follow his instructions

---

3*Ibid.,* #22, p. 251.
4*Kingston Gazette,* May 3, 1818.

from the Niagara Committee.[5] It had set June 6 as the day when the townships in this district were to hold the "training meetings" to elect the men who would meet on June 13 to elect delegates to the convention in York on July 6. On May 20 he set off for Ernesttown accompanied by Daniel Washburn in whose legal office former Massachusetts senator Barnabas Bidwell was manager and son Marshall was apprentice.[6] Shortly after Gourlay and Washburn had checked into the inn at Ernesttown, Daniel Hagerman called to see them. Gourlay accused Hagerman of removing property illegally from the Kingston post office. Lawyer Hagerman was not accustomed to having his actions questioned. He flew into a rage and squared off for a fight, but Gourlay refused to lose his temper. A bystander supported Gourlay's statement that Hagerman was in the wrong. Hagerman turned against the bystander too. Gourlay and his friends left, but long after they departed, Hagerman's jaws still quivered "with words of learned length and thundering sound".

Hagerman rushed off to Kingston and before the day was over he was distributing a circular justifying his withdrawal of the Adolphustown report on the grounds that Gourlay's activities were disturbing the peace. He then offered to attend any meeting that the people might call to explain his actions. When a second Adolphustown meeting was convened with Daniel Perry in the chair, it proceeded to draw up another report to take the place of the one Hagerman had removed.

Gourlay continued on his way to discover who else opposed his project and set them right about his motives. On May 22 he took the steamboat down the St. Lawrence. At Gananoque he stopped off to see Colonel Joel Stone who was also holding back a report.[7] Seventy-year old Joel Stone had lived through two wars with the United States and had spent three years in London from 1783 to 1786 seeking compensation from the king for his losses in the first one. He was now farmer, miller, road commissioner, collector of customs, colonel of the Second Regiment of the Leeds Militia and claimant of £1,300 compensation after the end of the war in 1814. Stone was not at home when Gourlay called, but when he saw him later he treated Gourlay with scorn. Stone had lost none of the anti-republican virulence that had caused him to come to Canada.

Gourlay proceeded to Brockville where he left some Niagara pamphlets for sale during the half hour boat-stop and then continued to Prescott, the

[5]*The Banished Briton*, #30, p. 405; *Narrative of a Journey through the Midland, Johnstown, Eastern and Ottawa Districts* contains the complete story.
[6]Barnabas Bidwell has been defended so vigorously that there seems little doubt of his innocence. J. Bethune wrote to John Macauley, Nov. 16, 1810; "The author of The Friend to Peace is supposed, here, to be a Mr. Bidwell, a gentleman who was obliged to fly his country for defrauding the public of their money . . . having taken 20 years to embezzel about 750 dolls!" (Macauley Papers, OA.) See also "The Barnabas Bidwell Case", by E. E. Horsey, 1941. (Queen's University Library, Kingston)
[7]*The Banished Briton*, #13, p. 119; #30, p. 406.

home ground of the "unlettered yeomen" led by Simpson. Through the columns of the *Kingston Gazette* he had set May 27 as the date for a meeting at the inn of Isaac Hurd. Only the inhabitants of Prescott were supposed to be there, but men were coming in from the whole country around. They crowded around Gourlay to shake his hand. One man slipped a note into his hand as he shook it. When Gourlay opened it, wondering at the secrecy, the note said that five to one at the meeting would be in his favour. When the meeting was convened, Gourlay tried to separate the local residents from the outsiders so that only the proper people could vote, but the disorder made it impossible. He called for a clerk to act as a secretary and collected money to pay the clerk. He attempted to establish parliamentary procedure for the discussion, but Jonas Jones, member of parliament for Grenville, lawyer from Brockville and former pupil of Strachan's, seized the floor. He harangued violently against the petition drawn up by the people of Niagara and the proposal to call a convention. As a member of the assembly that had not been able to get its own petition farther than Colonel Smith's desk, he was a dog-in-the-manger who could not allow another man to take up the task. It was in vain that Gourlay tried to explain that the purpose of this meeting was not to discuss the Niagara petition but only to elect a representative to go to the district meeting. When Jones would not give up the floor, Gourlay and his supporters left. Jones called Gourlay a wretch, a damnable wretch, and he published his opinion of Gourlay and his meeting in the *Gazette* on June 9.

Undaunted by this tumult, Gourlay continued next day to Johnstown (now Cornwall) where he was received cordially by Mr. Archibald McLean who wished well to the cause espoused by the people of the Niagara District. Gourlay gave him some of the pamphlets to circulate and next day set out by stagecoach for Lancaster. At every stop he pasted up placards with hours and places of meetings. By this time Philip Vankoughnett, a new member of the assembly and former pupil of Strachan, had seen a copy of the Niagara pamphlet and heard that Gourlay had just left to distribute his papers, pamphlets and placards along the road. He jumped into a little wagon and dashed off to catch this arch-enemy of Upper Canada. The paste on Gourlay's placards was barely dry when Vankoughnett clattered up to tear them down, heroically inviting loiterers to inform Mr. Gourlay that it was Philip Vankoughnett who had done the deed. If Vankoughnett counted on tiring Gourlay, he was wrong. Gourlay continued on his way, leaving his placards behind him.[8]

---

[8]*Ibid.*, #16, p. 152; *Statistical Account II*, p. 576. Philip Vankoughnett had commanded a battalion in the recent war. Later he commanded a battalion against Mackenzie. Twenty years after he rattled after Gourlay, he was still so anti-American that he threatened to fight any man who sang Yankee Doodle in his presence. His son became Chancellor of Upper Canada.

In two days he passed through Glengarry that had been settled by groups of disbanded Highland soldiers. Much of the land there was held by absentee merchants in Montreal waiting for higher prices. The people were poor. Before Gourlay paused at some of the houses, he learned to calculate the cleanliness of the refreshment he would be offered from the look of the windows, the piling of the wood or the washing on the line. The people were suspicious. Some thought he was a tax agent sent from Britain.[9] When he tried to get fresh horses, they made excuses. He proceeded with difficulty to the Ottawa River. In Hawkesbury Township Mr. MacDonell, M.P., and Mr. Mears, a former member, said they would help him by subscribing an amount proportionate to other districts, but because of the distance they would not send a delegate to the convention.

He had reached the eastern apogee of his hitching-post campaign through Upper Canada, the first man to travel through it for a political purpose. Few men in York, except Powell who had served on various law circuits, had traversed the whole province. On June 3 he began his journey back. Now he found that horses were cheerfully volunteered, and a supporter had called two training meetings for June 4 that he was able to attend. As he approached Cornwall, rumours began to drift to him about astonishing events since he had left it. Vankoughnett had been busy with more than tearing down placards. On June 4, the King's birthday, the militia paraded with flags and drums. The celebration ended with a climactic bonfire. The militia had bought up all the Niagara pamphlets they could, and after denouncing their distributor in orations in which they stated that the people of Upper Canada were blessed with a "mild and just government which watched over their interests with parental anxiety", they burned all the pamphlets in a pyre as high as the one that had burned Norman Gourlay three centuries before. When Gourlay arrived on June 5, the ashes of his pamphlets were still warm. Instead of being alarmed, he was contemptuous. When he met McLean on the street, he asked him if he meant to revive the barbarisms of two centuries back, but McLean was in no mood to discuss history.[10] This was the day on which the Cornwall meeting was scheduled to elect a representative. With unprecedented prudence, Gourlay hired a horse and rode off into the country around the Petit Nation River to see if it were suitable for the grand canal he was planning to circumnavigate the Cornwall rapids.[11] If he had stayed, his ears would have been blistered by the profanity of John Crysler, another member of the assembly, whose anger against Gourlay was as hot as the flames that had burned Gourlay's books.[12]

---

[9]*Statistical Account I*, p. 521.
[10]*Ibid.*, #26, p. 335.
[11]*Statistical Account I*, p. 577.
[12]*The Banished Briton*, #30, p. 409.

After the Cornwall episode, Gourlay returned to Kingston to allow the district to cool down and to replenish his supplies of campaign literature. At Prescott he took time to write to his wife. She had written to him that the Master of the Rolls had not accepted the Duke's latest lease, insisting that Gourlay was to have the kind of lease that he should have had at the beginning. He had been awarded £385 costs and interest, but it was still in the hands of the lawyers. The Duke's men had prevented Gourlay's employees from threshing the corn so that it could be sold to pay the rent. Her brother Thomas had washed his hands of the situation. He told his sister that if she would abandon Deptford Farm she and her five children could come home to him in Scotland.[13] Robert wrote her that no matter what happened she was to secure his papers, for they were the only proof of his good character. He told her that he had heard twice from Mr. Birkbeck whose colonization scheme in Ohio was meeting with success and knew that he could be doing as well if it were not for York's jealous system of peculation. Jean had been ill but her January 25 letter reassured him: "It has pleased God to remove my delicacy, and make me as strong as I ever shall be: and I hope it is not from presumption that I trace to the same high source the total want of fear which I experience towards our future destiny."

Robert answered with equal confidence from Prescott, June 7:

"The interest I have excited in the province by writing is very great, . . . I feel no blame in remaining here, . . . I have drawn for £60 on John Ranken, which I spoke to him about before leaving England. My expenses have been kept as much within bounds as possible; and I trust that these twelve months spent in Canada have not been thrown away. . . . I cannot imagine where you are, or what has happened, but have the utmost confidence that under every circumstance you will act with firmness and propriety. My hope of this amounts almost to pride. My attachment and regard grow with remembrance and with my experiences. Society is far from being in a right state here, either as it respects men, women, or children. I have seen strange scenes, but must say nothing of individuals till we meet. A most strange breaking up between me and your cousin has occurred, but I can give you no details sufficiently to inform you of all, and you must wait for them till we meet. I am conscious of having acted right, and shall satisfy you. For particular reasons, I wish you would write a letter immediately to your cousin, Robert Hamilton, at Queenston, signifying you having made over your third of the land, jointly owned by your mother and brother, to me. I wished to offer it to Mr. Clark

---

13*Ibid.*, #25, pp. 315-317.

as payment of my debt to him, but he would listen to no proposal and has publicly asserted that this is not mine, to injure my credit and ruin my character . . . I could have wished to send home a statement of its value, lest such might have been wanted by creditors; But I could obtain no estimate."

Then he added with melancholy truth, "It is, I believe, unsaleable."

As he returned to Kingston by boat he admired the incomparable scenery of the Thousand Islands.—"The Islands of every shape—rocky and tame—countless in their watery labyrinths, present variety infinite," and visioned ocean vessels from Egypt or India on the lakes.[14] On the evening of June 11, as he sat recounting the results of his tour to his friends in Walker's Hotel, a message was sent in to him that he was wanted outside the dining-room. It was the sheriff who handed him a warrant for his arrest on a charge of seditious libel. It was signed by Thomas Markland, J.P., directed by Attorney-General J. B. Robinson, through the agency of Mr. Christopher Hagerman on the complaint of Mr. Stephen Miles, printer and publisher of the *Kingston Gazette*. Gourlay had attended church with Miles only a few weeks before. Miles had accepted all Gourlay's communications freely and had noted with satisfaction the increased circulation that his paper was enjoying since Gourlay had arrived in the province. Now he accused Gourlay of being an enemy of the state.[15]

Gourlay could have been across the border in a trice with the assistance of his friends. Cobbett had fled from England but flight would not have served Gourlay's purpose. He had known for some time that he was courting arrest and he was willing to make himself a test case in Canada to ensure free discussion of public affairs. He went immediately to Markland to give himself up and to find out that the seditious libel for which he was being indicted was part of the draft petition to the Prince Regent. This petition had not been drawn up by him but by the Niagara Committee. It had not been approved for final presentation to the faraway throne but only as a basis for discussion. Nevertheless, Gourlay had been singled out to bear the brunt for all. Bail was set at £1,000, a sum so large that it was clear that his prosecutors intended him to be in jail till the August assizes in Kingston. An unnamed supporter put up the bail and much to the chagrin of his accusers, Gourlay walked away free.

Two people in York were busy claiming the credit for Gourlay's arrest. Administrator Smith's communication to Lord Bathurst of February 23 had barely reached its destination when he bundled up the Third Address

---

[14]*Ibid.*, p. 113.

[15]*Ibid.*, #30, p. 410. Miles had come from Vermont to Montreal in 1807. He and a partner transported a press by bateau to Kingston. His paper was the only one published in Upper Canada during the late war for the Newark (Niagara) and York presses had been seized by the Americans.

of April 18 and sent it to join the Second. "It is not possible to anticipate the result, but I am persuaded that any failure on the part of the Government to repress these measures would add fuel to the flames. I have however directed the Attorney General to watch the progress of this person and his employees in order to seize the first proper occasion to check by criminal prosecution the very threatening career now entered upon."[16] Colonel Smith had not thought it necessary to advise Dr. Strachan of his correspondence with London. Strachan was forced to learn about it through his friend Colonel Harvey, secretary to the governor-in-chief in Quebec. Strachan told Harvey that because Chief Justice Powell was in Boston he had been able to rouse Colonel Smith to the danger of Gourlay's meetings, and early in June a copy of the Niagara pamphlet reached Governor-in-chief Sherbrooke in Quebec "by private channels".[17] Strachan asserted that it was he who had directed the attorney general, John Beverley Robinson, to look over Gourlay's publications to see if an action could be brought against him, and the attorney general was only too glad to do it, for Strachan was his foster father. Any prosecution resulting from the attorney general's investigation, Strachan continued, must be undertaken while Powell was absent so that "a Judge on whom I have more dependence for impartiality will preside". Political impartiality had the same meaning then as now—complete concurrence with the views of the man who was making the statement.

In addition to these two men in York, a Niagara notable was indirectly involved in causing Gourlay's arrest. It was not Gourlay's relative Clark who had openly declared his disagreement with "the visionary enthusiast". It was Dickson, the one who had never stopped supporting Gourlay to his face, but who had been too canny to hand him the Niagara township report for which Gourlay had asked repeatedly, the only report he lacked to complete his survey of the Niagara District. Whenever he requested it, Dickson never seemed to have it ready.[18] Unlike Clark, Dickson had not committed to writing anything that might implicate him with the activities of the man who was swiftly losing favour with officialdom. If he should be accused of abetting Gourlay in what was now being termed a seditious path,

---

[16]Smith to Bathurst, York, April 18, 1818, Q324-31, p. 22.

[17]Strachan to Harvey, York, June 22, 1818, G. Spragge, ed., *The John Strachan Letter Book*, p. 163. Colonel Smith's character is a little hard to discern from the slim record. The indecision that history has ascribed to him may be due to Strachan's exaggeration of his own role at Smith's expense. On June 26, 1818, Powell wrote Gore in London that he was "no longer consulted by Mr. Smith. . . ." Gore had advised Powell, September 8, 1817, "to retire into your shell—these are not times for zealous and disinterested Public Servants". (Powell Papers, TPL.) The failure of Strachan's amendments to the District School Bill in both 1817 and 1818 was his first setback in influence since his provisional appointment to the executive council.

[18]*Statistical Account II*, p. 467. The position of legislative councillor carried with it the possibility of an hereditary title.

he would lose his position on the legislative council, according to the terms of the Seditious Alien Act of 1804. In addition, the Second Address had stepped hard on Dickson's toes when it called for an investigation into the sale of Indian land. Lawyer Dickson's fee for drawing up the papers in connection with the sale of the Grand River land was 4,000 acres, and later a choice 400 at the mouth of the river.[19] Even Dickson's brother Thomas had regarded it as outrageous. At that very moment Gourlay had in his hands a letter from some chiefs of the Grand River Indians asking him to intercede for them. An official investigation could be uncomfortable for Dickson.

There was another reason too for Dickson's change. Though the late session of parliament had spent most of its time, in Gourlay's words, in "flattering, fighting, and flying",[20] it had accomplished one thing before it absented itself from York. It had passed the fourth of Robert Nichol's resolutions that had caused Francis Gore to prorogue it the first time—the admission of American settlers. Americans were to be admitted as settlers under the 13th Geo. III and the 30th Geo. III. The comment of the *Niagara Spectator* was couched in London phraseology: "What effect will this have on the feelings of place-men and ministers, at York?"[21]

Now that he had obtained what was most important to him, Dickson could oppose Gourlay in the interests of preserving what York called the public peace. After his return to Niagara following the prorogation of parliament, he suggested to his Niagara neighbour, executive councillor Claus, that he arrest Mr. Gourlay. Colonel Claus declined the honour, telling Mr. Dickson that he could arrest Gourlay himself. This was the state of affairs in the Home and Niagara Districts while Gourlay was conducting his hitching-post campaign through the Eastern and Ottawa Districts.

The Midland District had not been idle either. Postmaster John Macauley had published a reply to Gourlay's criticism of his conduct as postmaster saying that he had done nothing wrong; as far as he knew, mail belonged to the person who mailed it.[22] Miles finished printing a pamphlet for free distribution entitled *Essay on Modern Reformers addressed to the people of Upper Canada to which is added a letter to Mr. Robert Gourlay*,

---

[19]Gourlay, *ibid.*, says that Dickson, through William Claus, Deputy Superintendent of Indian Affairs, was "allowed to buy" 6,000 acres. Powell cites 5,000 acres (Powell to Gore, York, 6 April, 1818, Powell Papers, TPL) which was payment for "perpetual legal advice to the tribes". C. M. Johnston, *The Valley of the Six Nations*, p. 180, says 4,000 acres. The choice 400 acres at the mouth of the river, referred to in Upper Canada Sundries, PAC, Jan. 5 and 30, 1818, seem to have been part of an acquisition before the war which the Indians were seeking to repossess.
[20]*Statistical Account II*, p. 545.
[21]The *Niagara Spectator*, June 11, 1818, copied from the *Upper Canada Phoenix*, June 2.
[22]*The Banished Briton*, #13, p. 115.

by John Simpson. Simpson classified three types of modern reformers, putting Gourlay in the third category of "self-created reformers . . . who court public favour by instilling the idea of imaginary wants, of visionary sufferings and fancied privations; who elevate political enthusiasm to treasonable mania; . . . We are ordered by the Mosaic law, not to set up any Brazen images to ourselves; how the good people of the Western District can reconcile their worship of you, with the commands of the decalogue I leave for them to determine". John Simpson praised the constitution, the merchants, religion and loyalty and damned Robert Gourlay for being a disturber. Pamphlet wars had a way of following Gourlay like dust eddies on the roads.

Though Miles had co-operated with the attorney general in swearing that the draft petition to the Prince Regent that he had printed on June 2 was seditious, he by no means refused the copy that Gourlay's campaign was pouring into his office.[23] He printed all the reports of the June 6 training meetings at Kingston, Fredericksburg, Adolphustown, Richmond, Hallowell, Ameliasburgh, Thurlow, Sidney, Loughborough, Marysburgh, Pittsburgh, Wolf Island, and the district meetings as well. Two days after Gourlay's arrest, the township representatives met in Kingston. When they found that some had made a mistake in the place and gone to Ernesttown, they all met again on June 20. They elected Washburn, Coleman, Peterson, Meyers and Hawley as district representatives to go to York. When Amos Ansley found that he was not one of the chosen five, he announced that he would go on his own to seek settlement of his boundary complaint. Only one training meeting, the one chaired by magistrate R. D. Fraser, said that they were much too happy to support Mr. Gourlay. The rest approved the draft petition to the Prince Regent and subscribed their dollars for the expenses of the delegates to go to London.

On June 12 Gourlay delivered to Miles for publication the first half of the report of his journey and prepared to retrace his steps to strengthen his support in the areas in which he had encountered most opposition. He despatched a man who could speak Dutch (Deutsch) into the country of Crysler and Vankoughnett. His agent reported that the enemy was losing ground daily. Gourlay himself went again to the Perth settlement. At this meeting he found good support, but unknown to him some one reported back to Colonel Joel Stone at Gananoque. On a torn fragment of paper he wrote on one side in small writing, "Gourlay's friends". On the other in larger writing was this:

> In favour of Gourley
> In the Township of Kitley

23*Ibid.*, #30, p. 405.

Benjamin Simon ? Esquir Chairman
Captn Dunkin Livinston Representative
Jesse Root his Sergeant was Clerk
Little Arnold ? one of the Committe & Timothy
Soper and the 3rd said to be William Brown
Captain Joseph Wiltsee said to be Chairman at
Wiltseetown at Dicksons Tavern[24]

On June 25 Gourlay was in Augusta again. When his agent went through to Glengarry, he had posted one of their 700 placards at Isaac Hurd's inn announcing a meeting for the same day. Gourlay spoke for the first three hours, asking for signatures in support of the draft petition. Jonas Jones then asked for signatures for a petition he had drawn up to present to the assembly. Gourlay scoffed that they could hope for an answer from the Prince Regent in four or five months, but as they had no idea when the assembly would be convened again, they might never hear an answer to Jones' petition. Jones' petition was as full of grievances as Gourlay's but his main objection to the petition to the Prince Regent was still, unaccountably, that there were no grievances to redress.[25]

Before this meeting was over, Gourlay had to ride off for Johnstown (Cornwall, the scene of the burning of his books) where he had made his next appointment. He had just alighted at the inn and was watering his horse with a bucket at the well when a man shoved one of his placards at him, asking if it were his. No sooner had he said "Yes" than a dozen people assaulted him from all sides, pounding him with fists and sticks. He dropped the bucket and got out of the scrap as well as he could with the help of a young man whose name he later learned was "Indian" Grant. His aggressors were still tumbling about the yard and crashing through the inn. Gourlay took some of the men into the shade of a tree to explain to them what he was trying to do for Upper Canada when the window above them opened and Duncan Fraser, J.P., bloody and profane, shouted at Gourlay to stop. Nothing short of a knockout blow could stop Gourlay if he had an audience, and Fraser and his party went off to find some other way of silencing him. Fraser came back with a fellow magistrate, John Mac-Donnell, M.P. for Prescott, before whom he had sworn that Gourlay was a seditious person, and they served him with a warrant for his arrest.

This was Gourlay's second arrest in two weeks, but this time he was not among friends. Magistrate Fraser demanded £3,000 bail, but Mac-Donnell, guessing that Gourlay would have little cash with him anyway, lowered it to £500. Gourlay had no alternative but to set off for Brockville jail in the custody of a constable. He was about to mount his horse and

24Colonel Stone Papers, Queen's University Library, Kingston.
25*The Banished Briton*, #30, p. 413.

proceed there with dignity when Fraser made him dismount and climb into the constable's wagon and move off as if he were in a tumbril. Again he was spared needless degradation by MacDonell who allowed him to ride his horse beside the constable in the wagon. Half-way to Prescott they met two men to whom Gourlay was known from the Augusta meeting. Outraged at his arrest, they put up bail for him. Once again, Gourlay was free.[26]

Gourlay knew his rights under the law. He retaliated by laying a charge of assault against Fraser. In one of the reports Gourlay had received at Queenston the previous December, the answer to the thirty-first question about what retarded the country was "a parcel of drunken magistrates". Gourlay had just encountered two of them himself. Now the same constable who had arrested him was despatched to arrest Fraser, but Fraser was in his own country. Hearing that the constable was coming after him with a warrant that would have forced him to appear at the same Brockville assizes as Gourlay, Fraser had his fellow magistrates allow him to appear before them at Quarter Sessions, not at the regular assizes before a jury of independent men. The young man who had come to Gourlay's rescue was fined £5 and sentenced to a month in prison, but Fraser was let off with forty shillings.[27]

Next morning Gourlay was off to his appointment at Elizabethtown. Here too the opposition announced a meeting at the same time as his. Colonel Sherwood, another Brockville lawyer, claimed that the people could petition the Prince Regent only through the assembly. He could not prove his point but he would not admit he was wrong. Two of Gourlay's other training meetings followed and then a general meeting of township representatives was held to elect the delegates for the great Convention. Next day Gourlay set off for Kingston where he called immediately for back copies of the *Kingston Gazette* to see what had happened during his absence.

Daniel Hagerman had answered Gourlay's letter which referred to his quivering legal jaw. He said he had gone to see Gourlay on May 20 "led by the same feeling of curiosity to see you, that we country people are generally actuated by when we hear of a wild beast being in the neighbourhood for a show. I had heard that you were from the Highlands of Scotland, and knowing that many of the productions of that country were considered very 'odd-fish', I felt a curiosity to see one that had been so much extolled by his keepers, and had become so noted for his growling

26Gourlay's quirks sometimes outstripped his magnanamity. He did not name the men who aided him. The *Kingston Gazette*, July 28, 1818, says that Gourlay met Zeba M. Phillips and David D. Jones, but there is no way of knowing if they put up the bail. Jones had been Fraser's prisoner during the late war on suspicion of giving information to the enemy. MacDonell had lost an arm in the War of 1812.
27*The Banished Briton*, #12, p. 108.

qualities. I was, however, much disappointed, for instead of discovering anything like austerity in your countenance, or seeing you command respect by the dignity of your demeanour, I saw you courting the applause of the spectators by a disgusting affectation of low familiarity." Hagerman said that Gourlay was a dolt's head on the matter of the post office, and his jaws would yet quiver to Gourlay's damnation.[28]

Gourlay now took the second half of the report of his journey through the Midland, Johnstown, Eastern and Ottawa Districts to the *Kingston Gazette* to be published. In it he claimed that there was a good possibility that his first arrest might have been engineered by Daniel's elder brother, Christopher, acting "without any positive authority from the Attorney-General. This Hagerman is the brother of the petulant fellow who cut such a figure at Ernest-town, and of another, who, I am told, was many years confined in the States' prison, for forgery, now reported to be hanged." Someone in the *Gazette* office read this report while the type was being set and rushed off to tell the Hagermans. Both men came storming furiously into the office. They urged Gourlay to withdraw the report about their brother's forgery before publication, asking him to remember their mother and sisters whose feelings must be spared. Gourlay replied that he had not only a mother and sisters but a wife and children, and that the Hagermans had no concern for them when they published their slander against him. Daniel then abandoned his coarse language and entreated Gourlay more civilly not to permit the publication. Gourlay said he would if Daniel would apologize for what he had said about him. "Hagerman stepped backwards,—swore a dreadful oath, and making a show of sawing across his shoulder with his hand, declared, that he would rather cut off his arm than make an apology." When the paper was published, including the allegation of forgery, a friend warned Gourlay not to leave the hotel without being armed as the Hagerman brothers were out to get him.

After dinner Gourlay walked out as usual with three or four friends but soon heard some one come bawling up to him from behind. Turning around, he saw Christopher Hagerman armed with a large whip. Gourlay's friends jumped aside and he alone bore the pain of the whip lashes Hagerman gave him. Then Gourlay taunted Christopher for hitting an unarmed man. A whip was thrust into Gourlay's hands, and the man who hated violence and had not struck a blow since the tussles of his boyhood found himself armed in a street brawl. Hagerman had taken care to attack Gourlay in front of the office of a magistrate friend. The magistrate seized Gourlay by the arm and demanded that he keep the peace. A bystander shouted that Hagerman was the aggressor but the magistrate would take no steps against him. Gourlay ordered Hagerman's arrest for

---

[28]*Ibid.*, #30, p. 424.

assault but Hagerman was released after he swore to keep the peace for a week.

The columns of the *Kingston Gazette* erupted with a series of letters from Publicola, Philo-Gourlay, Byestander, Junius. Fraser tried to spread the altercation to the columns of the *Montreal Herald* as well, but the *Herald* refused to publish his letter saying, "we have nothing to do with Village boxing, and club warfare, which never can be honorable to any community. They are a contempt and mockery of the law of the land."

The time would have been sorry indeed if no one could have seen the humour in this stark battle of opposites. On June 11, the eventful day in Kingston that witnessed the publication of Simpson's *Essay on Modern Reformers* and Gourlay's arrest for seditious libel, the "Postboy" was revealing some of his tender thoughts about a lady named Mrs. Eliza Matilda Freebottom to the "Traveller" (Dr. Howison) in the columns of the *Niagara Spectator*. "The Postboy gets beside you occasionally, my dear fellow Traveller, and was half inclined to have stopped you for a few minutes . . ." The Postboy was having a bit of fun with Gourlay's second "Address to the Resident Land Owners of Upper Canada" and those who were keeping "the press hot night and day. . . . All this and more I told Mrs. Freebottom. . . . I forgot to mention to you, my dear fellow Traveller, that Mrs. Freebottom is a widow, in the prime of life—and I believe in my heart, if farmer Gourlay was a single man, she would . . . . . . I can't say what, for she is wonderfully taken with him; she thinks he promises a great deal in too short a time, yet she has faith; she is pleased beyond measure, to think her house is to be filled and furnished from home, in ample order. Poor Kitty, the maid, will find a difference then, I take it,—the house will have to be swept at least six times a day."

The following issue burst forth with an answer from the indignant Mrs. Eliza Matilda Freebottom, who wished to cancel her subscription because the *Spectator* had exposed her tender feelings for Farmer Gourlay. ". . . You must not suppose, Mr. Editor, from what the Postboy has said, that I am a fickle maid in my 'teens, and so deeply smitten by Mr. Gourlay and his measures as to run headlong into destruction, without giving them one moment's reflection. . . . But I am willing to believe that Mr. Gourlay's plans will be the means of discovering from whence some of them originate, if the people of this province will awaken to their sense of duty, by coming forward in support of a Petition to His Royal Highness the Prince Regent." By now it was clear that coy Mrs. Freebottom was Upper Canada, and Kitty, the maid, was parliament, unwilling to undertake tasks that might cause her more work.[29]

---

[29]The end of this flirtation will never be known as the next six numbers of the paper are missing.

During the last six weeks, a new editor named Bartimus Ferguson had taken over the *Niagara Spectator* from Amos McKenney. He too knew the issues involved in Gourlay's struggle to goad parliament by means of the coming Convention. On June 30 Gourlay gave him for printing a little statement headed "Extracts from an opinion of Lord Erskine". Erskine was the great Scottish Whig jurist whose eloquent appeal to twelve Tory jurymen in 1794 had saved Thomas Hardy and his fellows from the death penalty for treason. "Erskine says, 'To create a national delegation, amongst a free people, already governed by representation, can never be, under all circumstances, a crime; . . . to constitute a legal charge of either of these offences, the Crown, as I before observed, must aver the criminal intention which is the essence of every crime;' . . ." This was accomplished by a footnote: "The case in which Lord Erskine gave this opinion was one of a delegation for petitioning Parliament. The reasoning is applicable to a similar delegation for the purpose of petitioning the King or Prince Regent. . . . for instances as the Petition proposed by Mr. Gourlay."

Below this insertion was a letter signed "Switch" saying that since parliament had no jurisdiction over the land-granting department why bother petitioning parliament? "Switch" and "Swing" were pseudonyms commonly used by English pamphleteers writing on behalf of the agricultural labourer. A large number of these so-called crude frontier people knew exactly why Gourlay was calling the Convention. The Convention was the equivalent of a motion of want-of-confidence before that useful and peaceful bit of parliamentary procedure had been invented. It would try to move a pachyderm called parliament. When the Convention met on July 6, they would know the extent of their success.

17

Convention of Friends to Enquiry

On July 2, 1818, Robert Gourlay once more strode up the gangplank of the *Frontenac*, this time on his way to York to supervise the Upper Canadian Convention of Friends to Enquiry. Little York was in a state of frenzy, for all its plans had been laid to have this incendiary behind bars during July so that a leaderless Convention would collapse. York was in a quandary too. Colonel Smith's appeals for guidance from Lord Bathurst of the Colonial Office had fallen on deaf ears. In May Bathurst answered Smith's February letter, but His Lordship made no reference to his plea for advice on dealing with the dangerous Mr. Gourlay. No answer at all came to Smith's April 18 communication containing the third address that called for a commission of inquiry. Lord Bathurst would occupy the post of secretary of state for the colonies for nearly ten more years, but time and again suppliants would come away saying that his only desire was not to be troubled. Smith must rely on local advice to cope with the man whom he considered a menace to the province.

On June 13, two days after Gourlay's arrest in Kingston, Smith received Attorney General John Beverley Robinson's written statement that Gourlay's publications contained "passages so plainly and so grossly libellous and so entirely subversive of that respect which the Government of every Country should vindicate to itself," that Robinson had no alternative but to order Gourlay's apprehension on his return to Kingston.[1] Smith also asked the opinion of the acting solicitor general, Henry John Boulton, brother of the man who had spread the news on the preceding October that Dr. Strachan was displeased with Gourlay's project for compiling a statistical account of the province. On June 15 Boulton replied: "the petition to the Prince Regent is I conceive a most dangerous publication and . . . is punishable as a libel. . . ."[2] Henry John's appointment as solicitor

---

[1] Robinson to Smith, York, June 13, 1818, Upper Canada Sundries, DA.
[2] Boulton to Smith, York, June 15, 1818, *ibid.*, vol. 39.

general was merely provisional because Lord Bathurst had refused to sanction his father's horse-trading in the shuffle of appointments following Chief Justice Scott's recent resignation. Attorney General D'Arcy Boulton had said that he would become judge if son Henry John could become solicitor general.[3] Bathurst had referred the matter to Sir Peregrine Maitland who would soon be embarking to take up his duties as the new lieutenant governor of Upper Canada, but meantime Henry John filled the position his father had provided for him.

Executive councillor Dr. Strachan's opinion does not seem to have been solicited, but it was despatched to Colonel Smith the day after Gourlay was arrested in Kingston on the warrant issued by Robinson. He sent it along Front Street, busy with the activities attendant on York's 800 people. He was determined that Smith should act decisively "when the safety and tranquillity of the Colony is at stake. Indeed no Gov't could stand were persons like Mr. Gourlay allowed to proceed without hindrance or opposition".[4] Strachan's main fear was that the publicity Gourlay was attracting would cause the British government to think that the colony was ready for rebellion instead of continuing in a state of prizing "their happiness in belonging to the most exalted nation upon the earth". He advised His Honour to transmit a special report on the proceedings of Gourlay and his supporters to London and to Quebec and to tell young Attorney General Robinson that he approved his actions. He also suggested that the administrator should dismiss all commissioners of the peace who approved Gourlay.

When the disconcerting news of Gourlay's release on bail reached York, the government leaders began to survey the possibility of curtailing his activities by declaring illegal the township meetings and impending Convention. On June 29 Robinson was forced to concede defeat. He had searched his books as far back as the thirteenth year of the reign of Charles II before he found an act passed against tumultuous petitions but as he found no record of any prosecution under that statute, he concluded that it would be unwise to interpret the law harshly against Gourlay. "The case is peculiar," he concluded, "and cannot be resolved by any precedent to be drawn from the Mother Country."[5] Robinson was not discouraged, however, for he assured Smith that Gourlay would undoubtedly be indicted when he was brought to trial in the autumn and if that too failed, he would lay a charge against the editor of the Niagara Spectator. On July 4 he reviewed all the arguments again and finally advised administrator Smith to seek the advice of the judges on the matter of

[3]Boulton to Smith, encl. Smith to Bathurst, York, July 7, 1817, CO 42/359/P191.
[4]Strachan to Smith, York, July 12, 1818, G. Spragge, ed., The John Strachan Letter Book, p. 169.
[5]Robinson to Smith, York, June 29, 1818, Upper Canada Sundries, PAC, Vol. 40.

the legality of Mr. Gourlay's activities in calling the convention. On July 7 Chief Justice Powell reported that none of Mr. Gourlay's actions violated the law.

The combined efforts of Smith, Robinson, Boulton and Strachan had proved ineffective. On July 6 fourteen out of fifteen of the representatives elected to the Convention of Friends to Enquiry gathered in York. Gourlay had organized his campaign to produce twenty-five, the same as the number of members in the assembly. In the districts where his writings had been circulated by newspapers the full number of representatives had been returned.[6] The Niagara District sent Robert Hamilton, John Clark, J.P., and Dr. Cyrus Sumner. Major William Robertson was absent through illness. The Gore District sent Richard Beasley, and William Chisholm of Nelson Township who at the age of sixty had fought at the battle of Queenston Heights. In spite of the arrests for libel and assault within its borders, the Midland District sent its complement of five: Daniel Washburn, Davis Hawley, Paul Peterson, Jacob Myers and Thomas Coleman. The Johnstown District that had burned the books sent Nathan Hickok. There were none from the Ottawa District that had not promised any and none from the Home District that contained York. There were only two from the Western and London Districts that had received Gourlay so cordially the previous December. Colonel Thomas Talbot had returned from London and descended on York early in June. No one but Colonel Talbot could call meetings in his territory. More people had wanted to attend the preliminary meetings than dared. One man who had a pension of £20 because of war wounds was afraid to appear lest he be deprived of his pension.[7] Nevertheless the Convention of Friends to Enquiry had mustered as many members as the number that customarily attended the legislative assembly.[8]

The fourteen representatives presented their credentials, called Richard Beasley to the chair, chose William Kerr secretary and Daniel Washburn assistant secretary. Gourlay was declared eligible to participate in discussions but not to vote for he was not an elected representative. As the great Convention deliberated, the guardians of the king's government trod the streets in anti-seditious flurry, ready to quell the violence they were certain these incendiaries would cause.[9] On two successive days the brother of provincial secretary Jarvis threatened Gourlay on the streets but Gourlay refused to be provoked to violence. On the second day of the Convention acting Solicitor General Henry Boulton distributed a circular

[6]*Ibid.*, July 4 and July 7.
[7]*Statistical Account II*, p. 576.
[8]The main account of the Convention is to be found in *Chronicles of Canada*, pp. 17-20.
[9]*Statistical Account I*, p. cciii.

inveighing against this assemblage of rebels.[10] This caused Amos Ansley from Kingston Township to issue a counter publication that declared vehemently, "We never ware Rebels and we Never will Be".[11]

Amos Ansley's farm was still in "peases" in spite of his petition in 1800, and in spite of support from others in the same district who suffered from the present uncertain boundaries. The Anglican divine, Reverend George Okill Stuart, estimated that his compensation should be £7,900. Some people, said Amos, had built barns and houses on property that others claimed belonged to them. Ansley had already sought the intervention of the law in the person of Thomas Markland, J.P., but his majesty's magistrate had done nothing against "the Act of a Rebellious Mob who laid violent hands on the Body of Amos Ansley, the said Ansley being in the Peace of God and the King alone, and in Quiet on the King's Highway in 1812 and committed him to prison without an oath and without a Trial. No eye to pity No Hand to Save". When he received no redress through the law, Ansley turned to the church and sought the help of the Reverend Mr. Stuart in abjuring his "neibours to Remove not the ancient LandMarks which they fathers Have Set Proverbe cXX v 20." When the church failed him, he attacked the "sleepy and Lazzy Priests that Nither Serve God nor the King".[12] When the attorney general, calling his petitions "very ridiculous papers", did not answer, he attended the township meetings, and when he was not elected a representative, decided to go to York anyway to give the Convention the benefit of his support in the hours after the sessions. The Upper Canadian government was as adept at tabling petitions as its British counterpart. Amos Ansley would petition graphically but unsuccessfully for ten more years.

As their discussions proceeded, there was enough altercation among the representatives to show that differences of opinion could arise even in a Convention chosen by the Elysian method of primary and secondary elections. An air of optimism pervaded the gathering as it discussed the probable political opinions of Upper Canada's new lieutenant governor, Sir Peregrine Maitland, and his father-in-law, the Duke of Richmond, who was coming at the same time to replace Sir John Sherbrooke who had resigned as governor-in-chief because of a recent stroke. Gourlay was certain that this Duke of Richmond was the well-known advocate of parliamentary reform who considered that every man who paid taxes should vote. Under Gourlay's moderate guidance, the Convention decided that it would not condemn Sir Peregrine before he arrived, but would give him a chance to prove that he could do better than his predecessor. The

[10]*The Banished Briton*, #12, p. 112. No copy of this seems to be extant.
[11]W. R. Riddell, "Humours of the Times of Robert Gourlay", *Transactions of the Royal Society of Canada*, 1920, vol. 14, p. 70.
[12]Kingston, August 2, 1818, Upper Canada Sundries, DA.

members adopted with few changes the petition to the Prince Regent that had been drafted in Niagara, but agreed to delay its despatch to the Prince Regent until they had presented one to Sir Peregrine. They drew up a new petition asking Sir Peregrine to call an election for a new parliament that would represent the will of the people better than the present one. Washburn and Kerr recorded a few objections to the petition, this petition was adopted and the Convention was adjourned, its business accomplished and the public peace still intact.

Dr. Strachan interpreted the failure of the Convention to despatch the petition immediately as a sign of weakness, not prudence. Mr. Gourlay had directed the representatives like children, he wrote to Quebec to Colonel Harvey who would communicate his point of view to the Duke of Richmond as soon as he arrived.[13] Dr. Strachan assured Harvey that the meetings "were watched so that nothing serious could have happened without prompt measures for we insisted upon being awake". He also told Harvey that he would get Colonel Coffin in York to help him secure the twelve hundred acres of land near Perth that Harvey had asked Strachan to help procure for him.

Dr. Strachan considered the situation to be more serious than an anonymous writer with the same sense of humour as the "Postboy" in the Niagara District. This wag published a "liebill" that went the round of the streets.

> "Resolved, that for the perfect security of the public money, collected for the defraying the expenses of the Commissioners to and from England, it be placed in the hands of our trusty friend, Barnabas Bidwell, Esquire; . . . Resolved, that Mr. Amos Ansley, as the most respectable in appearance of our body, be selected to present the petition to his Royal Highness, . . . And that before the said Amos Ansley proceed on his mission, a commission of Lunacy be appointed, to enquire whether there are any immediate symptoms of approaching madness. Resolved, that . . . be a committee to accompany the said Amos Ansley, and that they be particularly careful for the credit of the Representatives, that the said Amos Ansley do not run naked about the streets of London, blowing horns or trumpets as he has been occasionally wont to do. Resolved, that the Convention being rather short of grievances . . . any person . . . who will furnish them with any general grievance . . . or lie . . . shall be paid Twenty Dollars out of the Public Fund . . . and if the said particular lie shall concern the Reverend Dr. Strachan, they shall be paid five dollars additional— or any of his pupils, two dollars and a half. . . . Resolved, that Mr.

---

[13]Strachan to Col. Harvey, York, July 27, 1818, G. Spragge, ed., *The John Strachan Letter Book*, p. 171.

Gourlay shall be at liberty to make up a contingent account for plasters and bandages, and shall be allowed 3s 6d for every kicking, and 5s for every horsewhipping: . . . Resolved, that it is a grievance that there are not more grievances. . . . Resolved . . . that Mr. Gourlay's letters from his dear wife be forwarded by Special Messenger once a week, and that henceforth to preserve purity of morals and decency, order and decorum throughout the province, Mr. Robt Gourlay have absolute control over the Press, that nobody's lies and scurrility be published but his own."[14]

Across the sea, The *London Courier* laughed at the fuss the colony was making over the calling of the Convention for it regarded it as only part of the movement for parliamentary reform, a sign of healthy thinking.[15] The *Courier*, naturally, was left wing. Right wing Canadian leaders, though they would have a better laugh than Gourlay out of the "liebill", regarded it as their duty at all costs to prevent anything that might turn into a popular movement. Robinson had been studying law in London when the Prince Regent's carriage had been stoned as he was on his way to open parliament. Robinson feared anti-government demonstrations. He regarded the township meetings and Convention as "dangerous to this country, chiefly from their example, as they point out the mode by which popular movements on pretences less specious than the present, can be effected, and as we have no adequate military force in this Province, which it has been often found necessary to resort to in England to check the tumults excited by artful & discontented demagogues".[16] He considered it "seditious and illegal in the people to pass by their own proper representatives and their governor, and transmit an immediate remonstrance on public matters to the King, or the British parliament".

The Convention had come and gone with no violence except on the part of the government supporters, but Colonel Smith was still convinced that somewhere there must be rebels planning secret violence. Hearing rumours of strange doings around the mouth of the Credit River, twelve miles west along Lake Ontario, he sent Isaac Swayze to investigate. Spry, swarthy, scar-faced, five foot seven, sixty-seven year old Swayze had never reported any more money bags stolen nor uttered unguarded words.[17] To prove his loyalty, he had named his first son Francis Gore Swayze and his second William Dickson Swayze. Isaac and his brother Israel had been Quakers in the United States but soon, when electioneering in the Niagara District, Isaac would say slyly, "I pray with the Methodists."[18] Isaac

[14]Riddell Library, Osgoode Hall, Toronto.
[15]*London Courier*, July 8, 1818; *The Banished Briton*, #16, p. 167.
[16]Robinson to Smith, York, June 29, 1818, Upper Canada Sundries, Vol. 40.
[17]Letter from Mrs. Roy Summers, Fonthill, Ontario, Sept. 27, 1965.
[18]J.C. Dent, *The Story of the Upper Canadian Rebellion*, I, p. 27.

compensated himself for loss of esteem among his relatives by a gain in gov-
vernment circles. When the assembly had moved in 1817 that Lieutenant
Governor Gore be interrogated about the disbursement of the £2,500 civil
list, Swayze voted against calling him to account. He also voted against
the proposed taxation of wild lands that was directed against absentee
landlords, many of whom were government officials. Swayze was a depend-
able government man, ardent hater of republicanism. Because of his
experience with Butler's Rangers as a spy in the American Revolutionary
War, he was the right person to send to investigate a rumour. On July 20
Swayze reported to Smith that he had reconnoitred as far as thirty-five
miles along Dundas Street. The only suspicious activity he could find was
that some people would bake a barrel and a half of flour into bread and
disappear with it. Some said that they were working on a silver mine at
the head of the "Credot" and were sending the silver off to the States in a
small vessel that was sometimes seen coming and going from the lakeshore
at night. Others said that they were counterfeiting money. Nobody thought
that they were plotting against the government.[19] Swayze returned to
Niagara where he was hindering the sale of Gourlay's publications and was
keeping an eye on attendance at the continuing meetings that the township
representatives were holding to ensure that the government would know
that its actions were being observed.

The Convention at York had not been the end of the meetings. A
permanent committee of management had been set up and on July 20 an
"Upper Branch" Convention representing the Niagara, Gore, London,
Western and Home Districts was held at Newton's Inn, Ancaster, with
Richard Beasley in the chair. A continuing committee met at St. Catharines
to polish the addresses to be sent to the new lieutenant governor and the
Prince Regent. On August 1 a "Lower Branch" Convention representing
the Midland, Johnstown and Newcastle Districts met at Kingston. If the
people's parliament could not meet, the people's committees could. They
could also answer the accusation of government supporters that their only
reason for meeting was to get more land. To this Robert Gourlay replied:
"A person by the name of Coffin has spread the report in this quarter; the
same Coffin, I presume, in which the arrears of militia pay lay so long
buried,"[20] and the same Colonel Nathaniel Coffin, adjutant general of
militia in York, who would be asked to look after Colonel Harvey's
application for land near Perth.

At home, Gourlay's wife was writing to the husband who was busy
calling a Convention in another country and who could not return till he
had stood trial at the fall assizes. On July 18, 1818, she was far from calm:

[19]Swayze to Smith, July 20, 1818, Upper Canada Sundries, DA.
[20]*The Banished Briton*, #30, p. 400.

"I suspect the York junto have already poisoned the minds of their employers, as to your proceedings in Canada. Both the Courier and the Sun have got paragraphs with respect to the discontent in Upper Canada, occasioned by the adoption of the principles, and following the advice of a Mr. Gourlay. If you do not hasten home with all speed, they will have you transformed into a rebel for anything I can see to prevent it."[21] At last she had decided to sublet the farm, but her decision was nullified by Sir Samuel Romilly's ruling that the subletting could not be valid till the signator of the lease signed it.[22] The signator was far away in Upper Canada. Again she wrote on August 11: "Why should you not write me? So many arrivals! paragraphs in every paper from New York, yet no letter to me! I endeavour to fancy every disappointment is a good omen, yet, nevertheless I weary sadly. . . . If you come soon now, my dear Gourlay, you will probably find all your family in tolerable health; but I have had a terrible illness, from which I am slowly making my escape."[23]

Why did not Gourlay give up the struggle and return to England? In addition to open opposition, there was the physical strain of travelling in all kinds of weather through all kinds of country. The mosquitoes had not been his only physical hardship since he had come to Upper Canada. He left a description of one journey. He was approaching York from the west along the lake road through Etobicoke past Colonel Smith's house,

"lonely and desolate. It had once been genteel and comfortable but was now going to decay. . . . Not a living thing was to be seen around! How different might it be, thought I, were a hundred industrious families compactly settled here out of the redundant population of England! The road was miserable. . . . My pony sunk over the pasterns, and got afraid with the rattling gravel: he shyed at every bush; and was as foolish as a loyal-mad magistrate, alarmed with sedition: he was absolutely provoking. The half-spun appellations of Sterne's nuns would have been lost upon him. I cursed Little York for it; spurred, remounted, dragged, remounted, and spurred twenty times over, losing five minutes of time for every step in advance. It was my anxious wish to get through the woods before dark; but the light was nearly gone before the gravel bank was cleared. There seemed but one path, which took to the left. It led me astray; I was lost; and there was nothing for it but to let my little horse take his own way. Abundant time was afforded for reflection on the wretched state of property, flung away on half-pay officers. Here was the head man of the province,

21*Ibid.,* #25, p. 322.
22*Ibid.,* p. 319.
23*Ibid.,* p. 322.

'born to blush unseen', without even a tolerable bridleway between him and the capital city, after more than twenty years possession of his domain!! The very gravel bed which caused me such a turmoil might have been a turnpike; but what can be done by a single hand? The President could do little with the axe and wheelbarrow himself; and half-pay could employ but few labourers at 3s 6d per day, with victuals and drink. After many a weary twist and turn, I found myself on the banks of the Humber where there was a house and boat. A most obliging person started from his bed to ferry me across the river; but the pony refused to swim. With directions to find a bridge near at hand, I was again set adrift, lost and forlorn! The bridge was at last found; but a third time, lost! was the word, and that, too, in the very purlieus of Little York; for even to the church of that poor, dirty, and benighted capital, there was nothing like a direct and well-made road! No less than seven hours were thus wasted in getting over as many miles!! The first improvement of every country should be the making of roads; and, after that, speeches from the throne may be patiently listened to."[24]

Mrs. Radcliffe herself could have done no better with that description, but Gourlay was not writing fiction. He was working to relieve the poor of England. His beloved Convention had balked at hearing his poor law pamphlets but on August 6 the *Niagara Spectator* printed his address "To the Labouring Poor of England"; "Fellow men, Your fathers were free. The blood of many ages has taught us, that it is not by war alone that a nation may be enslaved." There followed the text of his petition of May 1816 to the House of Commons signed by the ninety-five poor of Wily parish, the poor he hoped to bring to a better land. His reasons for continuing the struggle were plain.

Because of the poor he was also an accused man in the eyes of the law till he could clear his name of the charge of seditious libel, a charge that accused him of being as dangerous in peace time as a traitor in war. He must begin now to prepare his defence for he had decided to plead his own case at his coming trial in Kingston. He was not a trained lawyer but he had read law as he prepared himself for his duties as commissioner of supply in Fife. He had followed the latest reform case of author and publisher William Hone who had been influenced by the London Corresponding Society.[25] In 1817 Hone was prosecuted for publishing satires on the government of Britain such as *John Wilkes' Catechism*. At Hone's first trial on December 18, 1817, he argued his own case and was acquitted. Chief Justice Lord Ellenborough was angry at the lawyer who had allowed

---

[24]*Statistical Account II*, p. 541.
[25]*Address to the Jury*, p. 14.

Hone to slip out of his hands. On December 19 Ellenborough himself tried to secure Hone's conviction at a second trial on the basis of another publication. He too failed. Gourlay decided to conduct his own defence, like Hone. He would go to New York to consult better law books than were obtainable in Upper Canada, and to engage a shorthand reporter to record his trial as he had done in 1816 against the Duke of Somerset. The whole world would be able to judge if he were seditious or not.

He opened the *Upper Canada Gazette* and there he saw a notice that fall assizes in Kingston were called for in August, a month earlier than usual. If he had not seen the notice in the paper, he would have been absent in New York, unable to honour his bail.[26] Now he had no time to go to New York for law books, but as he caught the boat to Kingston he was already preparing his defence in his mind.

---

[26]Gourlay cited this as an instance of persecution. Actually a bill changing the date of Courts of General Quarter Sessions of the Peace for Johnstown District had been passed on Feb. 9, 1818, and assented to on April 1, 1818. *Jls. Leg. Ass.*, OA Report #9, pp. 437, 565.

18

Gourlay Forever!

When Gourlay arrived in Kingston early in August 1818, the town was in
a flurry of fanfare, for the new lieutenant governor of Upper Canada, Sir
Peregrine Maitland, was passing through on his way to the capital at
York.[1] The ladies trimmed their bonnets and the gentlemen their tactics
for vice-regality in the persons of Sir Peregrine and Lady Sarah. Gourlay
denied himself the pleasure of waiting upon Sir Peregrine till he was
cleared of the charge of sedition. Now he learned to his dismay that Maitland's father-in-law, the new governor-in-chief, was not the Richmond
who was favourable to reform.[2] That duke was dead and this was his
nephew, one of Wellington's unemployed generals, the man who had
invited the top brass to the Brussels ball celebrated by Byron in *Childe
Harold*. Peregrine Maitland had eloped with the duke's daughter, Lady
Sarah Lennox. When the duke was given his new post, his son-in-law Sir
Peregrine was assigned to accompany him as lieutenant governor of Upper
Canada.[3]

As soon as the duke and his son-in-law arrived in Quebec, they were
shown a copy of the reformers' Niagara pamphlet and John Simpson's
answering *Essay on Modern Reformers*. On August 11 Richmond reported
to Bathurst that he considered that "the violence of Mr. Gourlay and some
of his friends has done much mischief to his cause and that what appeared
to be rather serious will turn out of small consequence."[4] Nevertheless,
reports of events in the upper province hastened Sir Peregrine's departure
from Quebec. After only one day there, he and Lady Sarah set out on the
journey to Upper Canada, visiting settlements along the way. Like Gour-

---

[1]*The Banished Briton*, #25, p. 320. Sir Peregrine was the first titled holder of this office.
[2]*Ibid.*, #13, p. 122.
[3]Maitland's wife was also niece of Lord Bathurst's wife, a circumstance that caused Gore
    to comment to Powell on the "petticoat predominance at Downing Street". Gore to
    Powell, Oct. 30, 1817, Powell Papers, TPL.
[4]Richmond to Bathurst, Aug. 11, 1818.

lay, he was horrified at the condition of the settlers in Glengarry and their priest, Alexander MacDonell, informed him that great portions of land belonged to absentee landlords in Montreal. By the time Maitland arrived in York, he knew that his first visit should be to the land office. "The Land Council," he reported with vigour on August 10 to Bathurst, "seems to have been sleeping over an Office choked with applications. I shall go to them every day and intend keeping them to it, till the Office shall be cleared.

"They were in the habit of meeting nominally twice a week, and may not perhaps be too well pleased at this additional application."[5]

Anxiety of preparation threw Gourlay into a bilious fever during the week before his trial. His trial was scheduled last on the *nisi prius* docket for the grand jury experienced difficulty in finding a proper indictment. On Saturday August 15, 1818, the courtroom was packed when the case of the King versus Robert Gourlay was called. Judge William Campbell was on the Bench, the judge whom Strachan had designated as being more amenable to his views than Powell. The charge was seditious libel contained in a passage taken from the draft petition to the Prince Regent as printed in the pamphlet, *Principles and Proceedings of the Inhabitants of the District of Niagara for addressing his Royal Highness, the Prince Regent.*

They did not select as seditious the section that praised the ardour with which Canadians had defended their country in 1812, or the statement which said Canadians regretted that Britons were oppressed with heavy taxation at home partly because of the burden of the colonies. They selected a passage which told the truth about themselves:

"Permit the loyal subjects of his Majesty merely to say as much, at the present time, on this subject, as may induce your Royal Highness to order inquiry to be made.

"The lands of the Crown in Upper Canada are of immense extent, not only stretching far and wide into the wilderness, but scattered over the province, and intermixed with private property, already cultivated. The disposal of this land is left to Ministers at home, who are palpably ignorant of existing circumstances; and to a council of men resident in the province, who, it is believed, have long converted the trust reposed in them to purposes of selfishness. The scandalous abuses, in this department, came some years ago to such a pitch of monstrous magnitude, that the home Ministers wisely imposed restrictions on the Land Council of Upper Canada. These, however, have by no means removed the evil; and a system of patronage and favouritism, in the disposal of Crown lands, still exists, altogether

---

[5]Maitland to Bathurst, York, Aug. 19, 1818, CO42/361/115.

destructive of moral rectitude, and virtuous feeling, in the management of public affairs. Corruption, indeed, has reached such a height in this province, that it is thought no other part of the British empire witnesses the like; and it is vain to look for improvement till a radical change is effected. It matters not what characters fill situations of public trust at present:—all sink beneath the dignity of men—become vitiated and weak, as soon as they are placed within the vortex of destruction. Confusion on confusion has grown out of this unhappy system; and the very lands of the Crown, the giving away of which has created mischief and iniquity, have ultimately come to little value from abuse. The poor subjects of his Majesty, driven from home by distress, to whom portions of land are granted, can now find in the grant no benefit; and loyalists of the United Empire—the descendants of those who sacrificed their all in America, in behalf of British rule—men whose names were ordered on record for their virtuous adherence to your Royal Father,—the descendants of these men find now no favour in their destined rewards; nay, these rewards, when granted, have, in many cases, been rendered worse than nothing, for the legal rights in the enjoyment of them have been held at nought: their land has been rendered unsaleable, and, in some cases, only a source of distraction and care.

"Under this system of internal management, and weakened from other evil influences, Upper Canada now pines in comparative decay: discontent and poverty are experienced in a land supremely blessed with the gifts of nature: dread of arbitrary power wars, here, against the free exercise of reason and manly sentiment: Laws have been set aside: legislators have come into derision, and, contempt from the mother country seems fast gathering strength to disunite the people of Canada from their friends at home."[6]

Henry Boulton, acting solicitor general, was in charge of the case of the Crown against its critic. The actions of the members of the Convention of the Friends to Enquiry, he said in his address to the jury, were as contemptible at York as they were at home. He ridiculed Mr. Secord and Mr. Beasley, the president of the Convention. He said that Gourlay was attacking parliament so that he could get a seat if it were dissolved. He compared Gourlay's conduct with that of traitor Willcocks and hoped that Gourlay's fate would be the same. If the people continued to listen to Gourlay, he would overturn the constitution. He accused Gourlay of taking refuge behind a printer's desk, not having courage to come before a Court of Justice.[7]

---

[6]*Statistical Account I*, pp. 575-576.
[7]*Address to the Jury at Kingston Assizes, in the case of the King versus Robert Gourlay for libel*, August, 1818, p. 4.

Gourlay rose to answer. He reminded Mr Boulton that Mr. Beasley, the president of the Convention, had been Speaker of the assembly for two years, and that many of the present members of the assembly were in favour of the actions of the Convention. Furthermore, Boulton's own father could be impeached for neglect of duty when he had been attorney general. He had failed to prosecute young Jarvis for killing young Ridout in an illegal duel, and (whether Gourlay mentioned this fact is not known) Henry John himself was an accessory to the murder, for it was he, as one of the seconds, who had awarded Jarvis the extra shot that had killed eighteen-year-old Ridout.[8] By now every eye was on this man who dared to infer openly in a court what many had been saying secretly—that there was one law for the rich and another for the poor.

It is not to be expected that Gourlay would be any less malignant than the counsel for the Crown. "Gentlemen, I have no patience with the whole of his stuff—it is all infamous. It is a disgrace to the British government to have such a thing as this acting Solicitor General. . . ." He spoke against the absent Mr. Robinson as well. "Gentlemen, the Attorney General of this Province is but a stripling—the foster-child of a certain clerico-political school-master; and, we cannot suppose him yet weaned from the influence of early established authority, to say nothing of those still more powerful influences to which virtue is exposed in such a nest of iniquity as the Capital of Upper Canada. It is well known how active the school master has been in every quarter trying to thwart my projects; and how virulent his pupils have become, because their master has not been able to accomplish his ends. . . ."[9]

As for his being seditious, and causing other people to be the same, Gourlay knew that another man was being accused of sedition at these assizes for saying, "Damn the Parliament!"[10] In England an accuser would be hissed out of society for entering such a complaint and any jury would dismiss the charge with contempt. Certainly he had sold the Niagara pamphlet, but it was the business of the Crown Counsel to prove that he had criminal intentions in doing so. "Gentlemen of the jury, I . . . flatly deny the charge of bad intention . . . what is now contended for is not *my* honour and *my* right:—it is the honour and right of thousands of your

---

[8] *The Banished Briton*, #12, p. 112. Had Jarvis been tried and found guilty Boulton would have been an accessory. If there was no crime, there was no accessory.

[9] *Address to the Jury*, p. 5 ff.

[10] This may have been John Vincent, secretary of the Kingston township second report, *Statistical Account I*, p. 477, which described reserves as a putrid carcass. He was sentenced to two months' imprisonment and £10 fine. In a letter to Maitland, Nov. 23, 1819, from Kingston, he wrote: "Every one that knows me can testify I am far from a seditious person. The Jury on my trial believes I am not, & say they are sorry they did did not understand the full nature of the case. Two females know I never used the words, the witness swore to and after I went away he could not remember the words, and wanted them to remind him it was the rotten Borroughs I spoke of, and I have heard the members say they were not the representation of the people."

fellow subjects." Gourlay could further confound the counsel for the Crown. He could prove that he had not even written the petition that contained the libel of which he was being accused. He would now examine his witnesses.

At this point the Crown refused Gourlay the right to examine all the witnesses he had brought for his defence. He would be allowed to examine Mr. Clark and Mr. Wilkie, but not Mr. Robert Hamilton or Mr. William Kerr. "Call Mr. John Clark," said Gourlay.

"What is your name?   John Clark.

What is your age?   30 years.

Where do you reside, and how long have you resided in Upper Canada?   Township of Louth and District of Niagara. I have been all my life in Upper Canada.

Are you a Justice of the Peace?   Yes, I am.

Do you know this pamphlet, entitled *"Principles and Proceedings, &c?"*   I have seen it.

Is this your name attached to the Preliminary Address?   It is.

Was your name placed there with your will and desire?   It was.

Are you the same John Clark mentioned in the pamphlet—in the 9th page as Representative for the Inhabitants of the Township of Louth, —and in the 18th page as chairman of a meeting at St. Catharines, and in the 19th page as Representative for Niagara District and member of the Committee?   Yes, I am.

As a member of this Committee, and by desire of your constituents, did you join in ordering this pamphlet to be printed and published?   Yes.

Did you, in the same capacity, order payment to the printer for the same?   Yes.

In the same capacity, did you order sundry persons to carry the pamphlets into different Districts of the Province, for publication? Yes.

Did you so carry any of them yourself, and where?   Yes, in the District of Gore.

Did you authorize me to do the like through the Midland, Johnstown, Eastern, and Ottawa Districts?   Yes.

Was I paid my expenses for doing so; and, did you authorize the same?   Yes.

Do you know or conceive that some hundreds of people were ultimately concerned in publishing this pamphlet?   I do.

Can you draw any line of distinction between my conduct in publishing the pamphlet, and that of any one employed by the Committee to do so?   None.

After having attached your name to the preliminary Address, and

ordered its publication, did you shrink from any responsibility incurred?   None.
Do you think me equally responsible as to the printing and publishing the pamphlet as yourself, as one of the Committee?   I DO NOT.
... In your conversation with me, did you ever mark any thing discreditable to my character; or, indicative of disloyalty?   NEVER."

The first witness stepped down and James Wilkie, brother of artist David Wilkie, was called.

"What is your name?   James Wilkie, 34 Point Henry. Ordnance storekeeper.
... How long have you known me?   As long as I can recollect.
... Did my father possess a large land property in the County of Fife? He did.
Did my father bear a high character there, and were not all my connections respectable people?   Very much so indeed.
Do you remember of my being Commandant of a corps of Fife Volunteers in Fifeshire?   Yes.
Were you under my command in that corps?   I was.
Did you ever hear of any stain on my conduct or principles, as a loyal subject, either as a civilian, or, in his Majesty's service?   No.
Do you think that I should be at all likely, or disposed, to stir up faction, or do any act with a malicious intention?   Certainly NOT."

Gourlay then produced the original draft of the petition to the Prince Regent signed by Dr. Cyrus Sumner and Major William Robertson. He gave a definition of libel and cited the case given in Saunders' Reports of Lake versus the King. The present Lord Erskine, he said, had gained immortal honour in upsetting previous judgments on similar cases. Formerly it had been insisted that jurors on libel cases could give their verdict only as to the fact of publishing, but in the interpretation of the law, they were governed by the judge. Erskine and Charles James Fox had regarded this as an infringement on the liberty of the individual and of the press. They had introduced a law into parliament that in cases of libel jurors should be free to decide for themselves upon the whole matter in issue, upon both fact and law. He cited William Hone's case as well.

Robert Gourlay had expounded the law expertly as Acting Solicitor General Henry Boulton had not. He had proven that he was not a principal but an accessory in publishing the pamphlet in question. He had not harangued the jury, he had appealed to their reason. When he dealt with the absurdity of trying to accuse the Niagara committee of sedition he had

---

11*Address to the Jury*, p. 10; *The Banished Briton*, #16, p. 164.

drawn admiration from the courtroom by recounting a story from Addison in the *Spectator*:

"An honest Frenchman travelling to Paris, ran short of cash, and could not get on. As an expedient, he tied up three small parcels, filled with brick dust, and marked upon them, respectively,—poison for the King— poison for the Queen—poison for the Dauphin. . . . He was seized, on the suspicion of treasonable intentions, and sent off to Paris, in high style, as a state prisoner. . . . The King, Queen and Dauphin had a hearty laugh at the facetious and ingenious contrivance to get speedily to town.— Let no one try such a trick in Upper Canada, for after there is proof of innocence, still crime will be the order of the day and brick dust will be poison . . . the very man on whose oath I have been arraigned continues to sell the pamphlets complained of." The story caused a surge of laughter, and another rose when he asked: "Pray what would be thought if I was lying in jail, while a Commission from the Prince, sent out to Upper Canada, should find all the allegations of the Address to be 'as true as Holy Writ'?"[12]

He had disposed of the charge of sedition. Now he asked why a copy of the indictment had been withheld from him. He was told that it was the practice of the court. He called it an arbitrary action. How could any man prepare his defence if he did not know the accusation? Why had the jurors been chosen only from Kingston instead of the whole district? If that practice of limitation were continued, soon the limits of justice would be confined to twelve men. Why had the assizes been held a month earlier than usual? If he had not seen the notice of the change of date in the *Upper Canada Gazette*, he would have dishonoured his bail, for he had not been notified. Why, at the last Niagara assizes, had poor Angelique Pilotte been condemned to death without benefit of counsel for murdering her newborn illegitimate baby? He had protested to some Niagara lawyers but was told that the judges did not allow counsel to plead in some cases. Gourlay had published his opinion of the practice and had been congratulated by several lawyers for voicing the same opinion they held.[13]

His peroration was nearly over. "My fate then, Gentlemen, and that of this great question, which concerns the invaluable right of free petitioning, rests entirely with yourselves; . . . Think not, for a moment, that this is a common case, whoever may tell you so, to throw you from your guard,—

---

[12]*Address to the Jury*, p. 18.
[13]*The Banished Briton*, #21, p. 225. Angelique Pilotte, a Michilimackinac Indian, had been taken to France where she became pregnant by a British officer. On her return to Upper Canada she was sheltered by Miss Elizabeth Hamilton as her waiting woman. When she was delivered of a stillborn child, she left it in the field. In the case of murder and treason, pardon could come only from the Prince Regent. Powell had applied for a reprieve but, because of the distance involved, was uncertain if it would arrive in time. She was finally "allowed" to escape.

think not that it concerns only me. A verdict of acquittal will not only clear me of unwarrantable scandal and reproach:—it will establish for yourselves and fellow subjects, your most valuable constitutional privilege, now most wantonly and audaciously assailed. A verdict of condemnation on the contrary, must cast a stain on thousands; and as I said before, should you commit me to durance, your country's reputation and your country's freedom must also be imprisoned."[14]

Judge Campbell "delivered a learned and able address to the jury," and they filed out. In half an hour they returned with their verdict, "Not guilty!" The roof rang with instantaneous applause and Robert Gourlay was freed from the charge of seditious libel. His acquittal removed the guilt from all the loyal inhabitants of the province who had signed and circulated the petition. It seemed such a great victory that no one, because of their tremendous, child-like faith in the monarchy, mentioned an imponderable remnant of the case—whether His Brightonic Prince Regent would see the petition that had already been dispatched to Bathurst.

Robert Gourlay had faith that he would, and the Earl of Kellie, entertaining Fife voters with tidbits from his hothouse, was never happier than Gourlay when he was the guest of honour at the dinner that forty of his supporters gave him the Monday after his acquittal. The toasts rang out: "His Royal Highness the Prince Regent. May his ears be opened to the Petitions of his loyal subjects and his hands ready to redress their wrongs; His Excellency, Sir Peregrine Maitland, Lieutenant-governor of this Province. May his administration justify the high expectations formed of it, by promoting an Enquiry into the state of the Province . . . ; Robert Gourlay, whose honourable acquittal we commemorate. May his slanderers, assailants and prosecutors blush for their abuse of him."[15]

Sir Peregrine at the time was preparing to add a few comments on the political situation to his August 19 letter to Bathurst describing the land situation. "A man of the name of Gourlay, half Cobbett, and half Hunt, has been perplexing the Province. They have found a Bill of Indictment against him, for a libel against the Government. I have not very great confidence in the issue before a petty Jury. But I hope he will not at least escape a heavy fine for a libel on an Individual, which will cripple him."[16] Neither he nor his father-in-law had come to Canada as reformers. They were the spokesmen of the Tories who had appointed them. They had passed the "Gagging Laws" in time of peace. They tried bread rioters for high treason. They had boosted the price of a penny newspaper to five pence so that a poor man could not buy. Their chief opponents, the Whigs' were bewildered and slow. Lord Cochrane reported that he intended to

14*Address to the Jury*, p. 18.
15*Chronicles of Canada*, p. 29.
16Maitland to Bathurst, York, Aug. 19, 1818, CO/42/361.

bring in a bill to abolish sinecures but it was not ready. Neither was Mr. Curwen's plan for the reform of the poor laws. By February 1818, Henry Brougham had been handed so many petitions for parliamentary reform that he acknowledged that "he felt himself at a loss to know what to do with the petitions he had been requested to present to the House. . . ."[17] Though Brougham advocated a wider franchise and increased education, he would not go too far in his recommendations for reform for he feared the crescendo of demands from below. The following year he himself would upbraid the government for not stopping "that torrent of Blasphemy and sedition which had lately inundated the country before it had arrived at its present height."[18] Every cottage in some parts of the country, he charged, hid some seditious publication.

Even as Robert Gourlay was standing trial in Kingston, the rafters of Deptford Inn were ringing with political speeches as Methuen, Benett and William Pole-Tylney-Long-Wellesley once again fought for ascendancy. The election poll lasted nine riotous days. Tory John Benett of Pyt House was losing. He tried to steal votes promised to Methuen to bring up his piteous count by pleading that Methuen had no objection to parting with them as he already had more than he needed. They were all fighting in a kind of battle that had no rules, but some fought to eradicate wrong and some fought to preserve tranquillity. Every paper in 1818 had a reference to the plight of the cotton spinners of Manchester, the weavers in Scotland, or the farm labourers of the south of England, but in August 1818, Sir Peregrine Maitland in Upper Canada linked Robert Gourlay's name with that of Cobbett and Hunt, and hoped that Gourlay would receive a fine sharp enough to cure him of being an agitator. In Scotland in July 1818, Joseph Hume was returned to parliament for the burghs of Montrose, Brechin, Aberbrothock and Inverbarie, the first reform member in Scotland, but in Upper Canada Robert Gourlay was preparing to face a second trial on the charge of sedition.

On August 25, he was in Brockville to honour his bail and answer Richard D. Fraser's charge that he was a seditious person. Four days of the assizes had passed, and still no indictment had been found against him. The Grand Jury had made one mistake at Kingston and could not afford another. When Gourlay questioned Boulton, Boulton said he had not fixed a day yet, and perhaps the trial might not be brought on at these assizes at all. On the fourth day Gourlay demanded honourable discharge; he had already been acquitted in one court. Discharge was refused. The Grand Jury, two members of which were Fraser and MacDonell, was still racking its brains to prepare a foolproof charge.

17Quoted in *Upper Canada Gazette*, May 1, 1818.
18R. K. Webb, *The British Working Class Reader*, 1790-1848 (London, 1955), p. 46.

On Monday, August 31, they were ready. The indictment for libel was set out with a fierce preamble charging Gourlay with "diffusing discontents and jealousies, raising tumults", and then the fresh passages that had been chosen as libellous were announced. This time they were taken from the *Third Address to the Resident Landowners of Upper Canada* of April 2, 1818: first, "I had little hope of satisfaction from the sitting of parliament, after perusing the Administrator's speech from the throne; and this little was entirely extinguished with the disgusting reply made to that speech by your Representatives"; second, "It has been my fate to rest here nearly two months, viewing at a distance the scene of folly and confusion", and third, "the blessings of social compact are running to waste. For three years the laws have been thwarted, and set aside by executive power;—for three sessions have your legislators sat in Assembly, and given sanction to the monstrous,—the hideous abuse."[19]

The pleading on behalf of the Crown was opened by Jonas Jones, M.P. for Grenville. He made a great point of the fact that "the monstrous,— the hideous abuse" constituted a malicious defamation of the character of the members of the assembly. In his reply, Gourlay maintained that libel implied malicious intent. He gave a resumé of his reasons for coming to Canada and the principles that motivated his actions. He said that members of the Commons house of parliament were servants of the people, and "being prone to betray their trust, it was a rule founded in right reason, that great liberties should be allowed in keeping them, by exposure, to the strict performance of their duty:—..." He said that dozens of addresses similar to the one for which James Durand had been expelled from the assembly and similar to his own flew around every borough in England during a general election. He then attempted to call witnesses to prove that he had not caused tumults at meetings in the District of Johnstown, but the judge would not allow him.

The reply was made by Henry John Boulton, acting solicitor general, who said that the representatives at the Convention of Friends to Enquiry were as contemptible at York as they were at home and maligned in particular Mr. Beasley and Mr. Secord. He concluded with a eulogium of his own family and connections, the conduct of whom he defied any one to impeach. The jury retired and came back with the verdict: Not guilty. Once again a courthouse roof rang with shouts of "Gourlay and Freedom!" Henry John gathered up his papers and departed, beaten again.[20] Gourlay reported the course of this second trial that acquitted him in the *Kingston Gazette*, emphasizing the fact that had he not been personally assailed he would not have stooped to personalities. He regretted the tone he had been

19*Statistical Account II*, pp. 581-2.
20*The Banished Briton*, #12, p. 112.

forced to adopt except that by it he had roused the people of Upper Canada to action and had only lived up to the motto of his country, "Nemo me impune lacessit".[21]

There is no record of Gourlay's two trials in the minute book of the Midland Assizes, August 15 and August 31, 1818. The reason is to be found in a note written by a clerk whose wishes had been over-ridden by acting solicitor general Henry John Boulton. Clerk John Squire wrote in his new book, "The Minutes of Proceedings for the Eastern, Johnstown and and Midland Districts were taken away from the Crown Office by Henry John Boulton Esquire, Solicitor General, *personally*, and never returned, although often asked for, for want of which, they are not recorded." Witnefs—John Squire.[22] Perhaps the "Modern Reformer" was not so daft when he considered that certain actions of the Upper Canadian government should be investigated.

Once again Gourlay's supporters offered to hold a celebration dinner for him, but this time he declined. He was leaving immediately for New York to get his wife's letters, for in March he had written her to send his mail there for him to collect on his way home.[23] He was departing in triumph. Twice he had established the principle that Upper Canadians had the right to petition the throne in England and that his criticism of the government was not seditious. He had confounded his detractors. His name had even been linked with two illustrious countrymen. On June 15, 1818, a Glasgow friend now in the United States had published in the *Washington Intelligencer* a brief biography of the unknown man who had burst on the political scene like a meteor and whose actions were being reported in the press of three countries. Robert Gourlay, he concluded, "acts from real principle: there is not a particle of designing mischief lurking about him: in conversation he possesses a frank and honest zeal . . . and he is one of three school fellows who will do honour to Fifeshire;—the two others are Dr. Chalmers, the present famous Presbyterian minister of Glasgow, and David Wilkie, the Royal Academician, the no less famous painter of the day, in a style unique, and entirely his own."[24] This had been copied into the Canadian papers and its few small errors of fact were duly corrected later by the truth-loving Mr. Gourlay. This praise was now spreading over all Upper Canada, and when he left Kingston, fences and corner posts were blazoned with his name.[25] Gourlay Forever! they proclaimed.

21Wha daur meddle wi' me.
22OA.
23*The Banished Briton*, #12, p. 112.
24*Ibid.*, p. 106.
25*Ibid.*, p. 108.

19

Gagged, by Jingo!

At nine o'clock the morning after his second acquittal, Robert Gourlay was aboard the steamship on his way to New York via Montreal. At Cornwall, hearing that His Excellency, the Duke of Richmond, would pass through on a tour of his new domain, he sent him copies of his four Canadian pamphlets containing his recommendations for reform, not knowing that some one else had already furnished Richmond with them from a different motive.[1] He wrote Sir Peregrine asking for his patronage for his proposed statistical account of the province. At Albany he boarded the Chancellor Livingston and at noon of September 11, 1818, arrived in New York and the City Hotel. All his wife's letters except one written on June 7 had just been forwarded to Queenston.

Controlling his disappointment he walked out to look at the glorious prospect from the Battery. He asked two bystanders to name certain places that he was viewing. When they learned that he was from Canada, they immediately asked, "What is that Gourlay doing there now?" He replied that Mr. Gourlay was doing no harm at present, and then made himself known. The men hailed him as the Washington of Canada, for reports of both his trials had been given fully in the leading papers.[2] Henry Boulton, in his speech as crown counsel, had accused him of trying to induce Canada to join the United States. These bystanders were surprised when they learned that all Gourlay desired was better administration in order to prevent that happening.

Now he turned his attention to his own business and invested Andrew S. Garr with power of attorney to settle his affairs in England. He sent a report of the Convention and its proposed address to the Prince Regent to Lord Erskine in England and despatched the whole in duplicate in tin boxes by different ships. To Jean he wrote: "My dear Jean, count upon nothing as certain. I am tossed upon the capricious wave, and my destiny is

---

[1] *The Banished Briton*, #13, p. 120.
[2] *Ibid.*, p. 15.

beyond my direction. I trust God will give me strength of mind at least not to sink down in despair. . . . It is a heavy reflection with me that I must be for a time drawing upon my friends . . . if I live, surely I shall be able to discharge the loans."[3]

To Ranken, his man of business, he wrote: . . . "My being bred to no profession . . . is quite perplexing. Could I sustain myself in the meantime, the best plan would be to qualify for a Canadian lawyer, which requires no great depth; and with the popularity I have gained, employment would soon fill my hands. . . . If I could get into the good graces of some honest attorney in London, to give me bread and cheese . . . as a student in the inns of court, the project might be made out; but I do not imagine that a broken farmer could get this accomplished. The fact is, were I a boy of fifteen years old, I could get on: people would make allowances and assist. To a man of forty years, there is neither allowance nor assistance. There is pride in raising a boy: pleasure in looking down on him who has fallen. But is he low who can call a Provincial Convention?"

While he was troubled about his livelihood, his wife was writing him on September 7, concerned about the cause they both supported, making no mention of her own anxieties: "I have seen with wonderful composure, a report that you, my dear Gourlay, have been arrested and held to bail at Kingston: I have not the least doubt that it is true; but I hope it will turn out for your benefit in the end, and for that of the cause in which you have so zealously engaged. My chief dread is that you may be rendered unhappy by hearing of my illness, at a time when you cannot hasten home to join me. It is my most earnest prayer that you may remain firm in integrity, let them try you how they may; but for my sake, and that of the children, I implore you to be calm; to submit to our lengthened absence in reverence to the First Cause; and to look upon the immediate instruments with pity and forgiveness. If you have sufficient power over yourself to continue composedly and constitutionally your efforts for the good of Canada, without noticing the petty malice to your own person, it will have a noble and powerful effect on the cause: and I know this is the course that you will naturally follow, if you are not agitated by opposing duties. I am now well: your children are just as you would wish them to be. My mother continues as kind as possible; and we have a prospect of soon going to Scotland, by leaving a woman in charge of the house here, as if we were soon to return. Endeavour to make your mind easy with respect to all at home, and try to turn your activity to individual benefit, as well as to general good."[4]

All Robert could do now was to return to Upper Canada to obtain the patronage of the lieutenant governor for his Statistical Account. On the

---

3New York, Sept. 17, 1818, *ibid.,* #25, p. 319.
4Deptford Farm, Sept. 7, *ibid.,* p. 323.

way down the Hudson he had met a former college friend, Mr. Thomas Kettle Young of Georgia, who gave him letters of introduction to some of his acquaintances in the New England states. He decided to return to Upper Canada by making a circuit and after leaving New York on September 20, reached Hartford, Connecticut, on Thursday. He found Connecticut busy remodelling its constitution, a procedure they considered normal, not seditious. He reached Boston late Saturday, and Monday morning found him in the Athenaeum deep in the study of the causes of the American Revolution. He gave the librarian, Mr. Shaw, copies of his pamphlets to read. Mr. Shaw returned with astonishment to him. "I thought you were for us," he said. Again Gourlay answered, "No, I am for Canada."[5]

He continued to Salem and Cambridge and spent Sunday in the Shaker settlement at Watervliet. In their service he found the same deep reverence that he had seen in the Catholic service he had attended in Montreal on his way into the province. He also observed that it was the Methodists who were keeping alive the flame of religious worship in America as they had done in England. At Pittsfield he dined at the fall agricultural show where the chairman announced a toast to him and a member of Congress said that if he went to Washington he would be given a seat on the floor of the house like Lord Selkirk. The father of the future lieutenant governor of Massachusetts invited him home, and next day he judged the ploughing match. At the dinner that day he proposed a toast: "Commerce free and nations friends." The reception he was accorded was soothing after his treatment in Canada. Travelling to Sacket's Harbour, which he was pleased to visit when under no compulsion from Thomas Clark, he crossed Lake Ontario and was back in Kingston on October 17.

Nothing had changed in Upper Canada. A letter in the *Kingston Gazette* accused Gourlay of going to the States to obtain money to subjugate the province. Equally damaging to him was a tidbit of news that Thomas Clark had seen in the *Montreal Herald*, which had copied it from the *Albany Gleaner* of October 1, which had copied it from the *London Courier* of July 8. A reporter had written that the proceedings in Upper Canada "appear to have originated with a Mr. Gourlay, one of the worthies who escaped after the disgraceful proceedings of Spa-Fields".[6] Gourlay had stated that he had gone to Spa Fields only to observe the character of the meeting, but Clark accepted the *Courier's* report as truth and showed it to his friends in the government. During his absence also, editor Stephen Miles had concluded the publication of Gourlay's *Narrative Addressed to the Worthy Inhabitants of the District of Niagara*, the account of his journey through the Midland, Johnstown, Eastern and Ottawa

---

[5]*Ibid.*, #1, p. 16.
[6]*Ibid.*, #16, p. 158; #25, p. 324; *Statistical Account I*, ccxvi.

Districts. His readers had demanded it. Miles had also confessed that he was not the author of an article on July 7 derogatory to Gourlay. Mr. Pringle, a government friend of postmaster John Macauley, was the author, and threatened to cancel his subscription if Miles did not publish it. Miles now commended himself to God and promised to be neutral in future, but the damage to Gourlay had been done.[7]

The new lieutenant governor had convened parliament on October 12, 1818, for there was no money in the public treasury. In the Speech from the Throne Maitland deplored recent attempts to "excite discontent, and to organize sedition". He asked for a law to declare the township meetings illegal. As the assembly debated the Speech from the Throne, George Hamilton took notes to give to Gourlay. The member from Wentworth, James Durand, termed Gourlay's publications libellous but he said that his followers were honest, loyal, respectable men, better than many in the assembly itself. He said that the Convention was a natural growth from the desire of subjects to maintain their liberty but nevertheless he could not support it. Jonas Jones, the ultra Tory from the Cornwall district, called Gourlay a wretch, guilty of sedition and of organizing loyal men to the same. Another called Gourlay a great seducer, able to persuade the people to anything. Nevertheless, when this member had told his constituents that they had no grievances, they reminded him of the £3,000 gift for Gore. Colonel Burwell of the Talbot settlement said he knew Gourlay had evil intentions. Burwell wrenched out of context one of Gourlay's statements in the Third Address and said that "surely British blood, when it has ebbed to the lowest mark, will learn to flow again!"[8] It was clear that Gourlay had meant that Britons must revitalize themselves in order to preserve their liberty, but Burwell claimed it was incitement to rebellion. The township meetings must be stopped.

Gourlay was in Kingston when he received word of the acrimonious debate. Immediately, on October 20, he issued an address "To the Upper Canadian Friends to Enquiry" protesting the Speech from the Throne that labelled them all seditious. Again he called for township meetings. Eleven were held in short order around Kingston, asking Sir Peregrine to hold another election for the present parliament no longer had the confidence of the people.[9] Gourlay attended the meetings on his way to York to plead against the passage of the bill to forbid them. In the village of Amherst, now Cobourg, fifty people met on November 11. Gourlay spoke for three hours, and Charles Fothergill, the man Gourlay had met in York in October, 1817, spoke at equal length in a speech that he afterwards

[7]*The Banished Briton*, #30, p. 422; H. P. Gundy, *Early Printers and Printing in the Canadas* (Toronto, 1957), pp. 24ff.
[8]*Ibid.*, p. 581 for third *Address*; p. 665, for Burwell's interpretation.
[9]*The Banished Briton*, #15, p. 133; #16, p. 149.

printed in a pamphlet. "Look at this man", wrote Fothergill, "this self-created champion of liberty—narrowly mark all his proclamations, his unsteady, anxious eye, his hurried gait, his pockets welled out with combustibles, which he is ever ready to deal forth by handfuls in the hope that one, at least, will be accepted—his artful attempts to raise up a standard of discord . . . then say whether you would take such a man, for your organ, your champion, and your great political leader!"[10] This meeting adopted resolutions adverse to Gourlay. Foiled by Fothergill's opposition in this locality, Gourlay proceeded through sparsely settled land to York, declaring that the farmers of the district of Newcastle were isolated and uninformed.

At York the assembly was busy apologizing to Sir Peregrine for its lack of co-operation with administrator Smith during the preceding session. They had ordered expunged from the minutes all references to the recent struggle between the assembly and the councils that had resulted in their dismissal. He found too that Sir Peregrine had refused to receive the petition that the Convention had prepared. On October 24 the committee appointed in the summer, Mr. Hamilton, Mr. Beasley and Mr. Kerr, had asked for an audience. Maitland refused it, saying he would not receive a petition drafted by an unconstitutional meeting.[11]

Though Maitland had arrived in the province with some prejudice against the previous actions of the executive, he soon concurred with their views on the Convention and township meetings. When Maitland declared his opposition to them, the assembly gave him no trouble when it was asked to pass a bill "to prevent certain meetings within this province". Thirteen members voted for it and only one, Willett Casey of Lennox and Addington riding, voted against it. Even James Durand, who had skipped out of York two years before to avoid being jailed, voted to suppress the township meetings organized by Gourlay. On October 31, 1818, after it was pushed speedily through all its readings, the Act to Prevent Certain Meetings became law. The township meetings were said to encourage riot, tumult, disorder, and persons guilty of promoting or attending them would be guilty of high misdemeanour.[12] The assembly claimed that the right of an individual to petition king or parliament was still intact. Petitions were sent from Ernesttown with fifty-nine signators headed by Peter Perry, from Hallowel with 102 signators, and others from Cramaché and Percy deplor-

---

[10]C. Fothergill, *Proceedings at the Meeting of Hope and Hamilton Townships, Newcastle District*, York, 1818; *The Banished Briton*, #16, p. 149. Some people held the same opinion of Fothergill as he did of Gourlay. Dr. Thomas Rolph of York once told Mrs. Fothergill that her husband had every sense but good sense. J. Baillie, "Charles Fothergill, 1782-1840, CHR, December 1944, p. 384.

[11]*Ibid.*, #26, p. 344; Upper Canada Sundries, PAC., notes taken by Thomas Ridout.

[12]*Chronicles of Canada*, p. 31; Doughty and McArthur, *Documents relating to the Constitutional History of Canada*, 1791-1818, p. 554.

ing the continued persecution of a man who had twice been honorably acquitted and stating, "We must out of respect to our character as a people frankly say that our Confidence no longer rests with them and we do most fervently pray for a dissolution of the present Parliament that by another election of Representatives the sense of the People at large may better be known to your Excellency". Hagerman wrote to Maitland about the petitions—"signed by school children".[13]

Upper Canada now had its Gagging Act as well as Britain. As soon as the news of the final passing of this bill reached Niagara, Gourlay seized his pen. "Gagg'd," he wrote, "gagg'd, by Jingo!"

> "Dear sweet Canada! thou art gagg'd at last,
>   A babe of mighty Wellington, come o'er the sea,
>   Has, with thy own foul fingers, gagg'd thee!"[14]

Where was the power of the Commons now? he asked. All the quarrel about the power of the purse had gone up in smoke, and Sir Peregrine had announced that he himself would introduce in the next session a bill on roads. If the people's representatives could not even introduce their own laws, where was the power of the Commons? Where was democracy in Upper Canada?

Only the first instalment of this letter appeared in the *Niagara Spectator* on December 3 without the signature that would be printed at the conclusion. An unsigned letter was considered the responsibility of the editor. On December 16, editor Bartimus Ferguson was arrested for seditious libel on the accusation of Isaac Swayze on a warrant issued by William Dickson. As Ferguson was taken to the new Niagara District jail, Isaac Swayze swore jubilantly that in ten days Ferguson would have Robert Gourlay for company. Ferguson retaliated by writing his opinion of Swayze in the columns of his paper: "This man," he wrote, "while employed by our new Governor to supply his table with poultry, held a very suitable situation under Government; but, when sitting in Parliament, or galloping to Squire Dickson with information of libels and seditious publications; then, we say, he is quite out of the sphere of his understanding, and ought to be checked by timely chastisement."[15]

That same day Gourlay hastened to declare that Ferguson had been wrongfully jailed for it was he who was the author of the article. He went further and said that if the continuing committees were forbidden to call township meetings, they could convene as individuals. He asked those interested to come to St. Catharines on December 26 to protest the asser-

---

13Upper Canada Sundries, PAC, end of 1818; Jan. 6, 1819.
14*The Banished Briton*, #16, p. 145.
15*Niagara Spectator*, Dec. 17, 1818; *The Banished Briton*, #16, p. 153.

tions of the assembly that they were seditious. People could meet in towns, in counties, in districts, in Little York itself, to instruct their elected representatives how to act.[16]

Isaac Swayze, the guardian of the public weal, was jubilant now that Ferguson was in jail. On December 16 he reported happily to Major "Hillard", Sir Peregrine's secretary:

> "the buisiness is so far under way, that B. Ferguson one of the Editors is in close custidity and Benjamin Pawlin his partner, is held to bail for apperance at court, in the sum of £400 cury. The Honourables W. Dickson and W. Claws whom did ishued the Warrent to apprehend those persons I am sorry to say i see moore people in Government service, ready to assist the Editors of the Niagara Spectator than i really did expect to have found when last i saw you."[17]

On his way back to Niagara from York, Swayze had travelled overland by Dundas where his friend and fellow sleuth, Otaway Page, had his eye on a piece of property left vacant by a wartime deserter. Then he proceeded to nearby Ancaster where he ascertained that no township meeting had been held at George Hamilton's home but that Gourlay had been visited there by Colonel Beasley and Dr. Sumner. Thus far the government forces had been successful in their aim to gag free discussion of public affairs.

Swayze, however, had rejoiced a little too soon over Ferguson's incarceration. Dickson had not had the authority to swear the indictment against editor Ferguson, for he had not been a commissioner of the peace since April 1817 when Gore had relieved him of that office because he had persisted in administering the oath of allegiance to immigrating Americans. Ferguson was released and Upper Canada had an uncensored press again. The government forces had recently taken over the *Kingston Gazette* which was now in the reliable hands of Postmaster John Macauley and his friend Pringle. Miles was returning to the ministry, exhausted by the unexpected turmoil in the newspaper field.

The government of Upper Canada was pitted against an antagonist who had been observing politics for years and knew its slippery ways as well as they. Since township meetings had been declared illegal, Gourlay would call meetings under another name. The subsequent "purse-string" meetings were called to exhort the assembly to continue its fight for control of finances. Niagara, Grantham and Louth townships met at St. Catharines on December 26 for the purpose of instructing their member of parliament

---

16The instruction of elected representatives from below rather than above is a complicated issue. An able discussion of it may be found in B. Kemp, *King and Commons, 1660-1832*, pp. 40-50. The opposition cited mediaeval precedent but the government regarded it as a concomitant of revolution.

17Swayze to Hillier, Niagara, Dec. 16, 1818, Upper Canada Sundries, PAC.

about his actions. The officials were prepared for such cunning. They had already considered how to get rid of Gourlay if "An Act To Prevent Certain Meetings" failed to accomplish its purpose. On November 10, the day on which Attorney General John Beverley Robinson found no legal objection to "An Act to Prevent Certain Meetings", Chief Justice Powell, Judge William Campbell and Judge D'Arcy Boulton signed a paper (in Powell's handwriting) interpreting certain portions of the Seditious Alien Act of 1804.[18] This was the Act that had caused Gourlay to laugh to Street a few months before: "That, sir, is applicable only to aliens. I am a Roman. . . ." As the Act was popularly known as the Alien Act, there was justification for his laughter. Now the three judges declared that the oath of allegiance must have been taken in the province of Upper Canada and that the word "inhabitant" would be used in a sense other than its common one. For the purposes of this Act, an inhabitant would become an alien if he had been absent in a foreign country, leaving no fixed residence behind him, during the six months prior to a possible arraignment under the Seditious Alien Act.

According to the terms of the Act, the only men who could put it into effect were the lieutenant governor, judges, executive and legislative councillors. William Dickson, the man who had declared a short time before that he would live under the American flag if the government did not stir itself, was a legislative councillor who was willing to put the Act into motion. First, he must find some one to swear that Gourlay was not an inhabitant of Upper Canada according to the re-interpretation of the Act. On December 18, 1818, Dickson met with Isaac Swayze.[19] Isaac Swayze was asked to swear that Robert Gourlay was not an inhabitant of Upper Canada, though every one knew he had been living there for a year and a half. On December 19, 1818, Thomas Meritt, sheriff of Niagara, served Gourlay with a warrant for his arrest on the grounds that he was not an inhabitant of Upper Canada for he had been absent from the country for the last six months leaving no fixed residence there; and that he was an evil-minded and seditious person, who intended "to alienate the minds of His Majesty's subjects in this province . . . from his Person and Government".[20] If he did not leave the province before January 1, he would suffer death as a felon. The charge was laid by legislative councillor Dickson and executive councillor Claus on the oath that Isaac Swayze swore on the Bible that Gourlay had not been an inhabitant of Upper Canada for six months prior to the issuing of the warrant. His trip to New York to get his wife's letters was being used for his damnation.

18*Ibid.*, Nov. 10, 1818.
19*Statistical Account I*, lxviii.
20The original warrant is in the Ontario Archives; text in *Statistical Account I*, pp. xxvi ff.

Gourlay refused to leave on the grounds that he was an alien. He had been born in Fife and had taken the oath of allegiance in Britain; otherwise he could not have commanded His Majesty's forces there. Upper Canada never required the oath except for holding public office or obtaining land. He had published correspondence stating he was ready to take the oath in Upper Canada as soon as he was granted land. If he left, anyone would be liable for prosecution on the same arbitrary grounds. The Alien Bill had never before been invoked against a British subject.[21] If he left, he would forfeit "his sacred right of trial by jury". This threat to liberty was uppermost in his mind as he wrote about this latest development to his wife from Queenston, December 23, 1818.[22] When he refused to leave, he was taken to Niagara jail. Two days later (December 21, 1818) he was brought before William Dickson, Thomas Clark, William Claus, Alexander MacDonell, and Dr. Muirhead for questioning.

> "Do you know Mr. Cobbett? Yes.
> Do you know Mr. Hunt? Yes.
> Were you at Spa Fields meeting? Yes.
> Were you ever in Ireland? Yes.
> Were you lately in the Lower Province? Yes.
> Were you lately in the United States? Yes.
> Was it you that wrote the article in the Spectator, headed by 'Gagged, gagged, by Jingo'? Oh, to be sure it was!"[23]

After this interrogation, Dickson swelled with pride at having proved the danger latent in this man. "Gentlemen," he said to his colleagues, "Gentlemen, it is my opinion that Mr. Gourlay is a man of desperate fortune, and will stick at nothing to raise insurrection in this province." When Gourlay told him that he knew very well that he had been a resident in the province for more than a year, Dickson said that his name was not on the assessment roll, but the Act did not mention that criterion for an inhabitant. Supported by the other magistrates, Dickson ordered Gourlay to be taken from the courtroom into one of the prison cells to await judgment. After an hour he was brought back, and this little Star Chamber of Upper Canada gave Gourlay a written order to leave the province by January 1 under pain of death if he refused.[24]

21*The Banished Briton*, #26, p. 347. It had been applied twice against Americans during the War of 1812. A. Dunham, *Political Unrest in Upper Canada*, (Longmans Green, 1927), p. 50.

22*The Banished Briton*, #25, p. 324.

23*Ibid.*, #16, p. 165; *Statistical Account I*, ccxv.

24Under the Stuarts the Star Chamber became identified with personal despotism, the High Church party, attempts to abolish parliament, and the privilege of the king's friends over the common subjects.

Again he refused to leave. Twice he had been acquitted of a charge of sedition by a jury in a proper court. What authority had these individuals to declare him a seditious alien and order a British subject to leave a British colony? Did they think he was a simple oaf from the fields brought into a duke's kitchen to be sentenced on the word of a gamekeeper? Did they think he was like a young Scot who had been imprisoned for a petty crime and who on the day appointed for his execution had barricaded the door of his cell? The boy's mother (good brain-washed member of the lower class) called through the door, "Johnny, coom out and be hanged and dinna angre the laird".[25] Did they think he was in the category of the horse thief who had just been sentenced to death and then had his sentence commuted to banishment by Sir Peregrine? The horse thief was guilty. He was innocent, for no crime had been lawfully proven against him. Why should he submit to a violation of the liberty of the individual?

The law gave him the customary three days of grace to obey an order. Still he refused to leave. On January 4, 1819, sheriff Merritt came to take him to jail. The ten-inch wooden door was locked behind him and Gourlay faced the walls of the cell that would be his domicile till the next assizes in August.

His friends gathered at the jail, angry at the enormity of the law under which he was imprisoned. They said they would tear down the jail and free him, but Gourlay believed in non-violent methods. If he escaped, he said, nothing would be proved except that the Bill of Rights was a Bill of Wrongs.[26] His friends drafted two petitions on his behalf. Peter Hamilton and Alexander Hamilton, J.P., swore that Robert Gourlay was respected and esteemed in both Britain and the Niagara District. Robert Hamilton, eldest son of the Honourable Robert Hamilton, and James Kerby, J.P., swore that Gourlay had been domiciled in Queenston for more than nine months. Robert Gourlay swore that he was by birth a British subject and an inhabitant of Upper Canada for more than a year before the warrant had been issued. These three petitions were taken to York and presented with a request for a writ of habeas corpus. On January 20, 1819, after a little delay, Chief Justice Powell ordered that the body of Robert Gourlay, then detained in prison, should be brought before him at York.

On February 4, Gourlay set off with sheriff Merritt for the overland trip to York. On the 8th, Chief Justice Powell asked him one question: "Have you brought any one with you?"—meaning a lawyer, it is presumed. According to legal practice, it was possible to give Gourlay an immediate hearing before the Court of King's Bench. It was possible, but not probable, for such a procedure would have drawn more attention to Gourlay's case than

---

[25]*Statistical Account II*, p. 654.
[26]*Appeal*, p. xliii; *Statistical Account I*, ix.

was desirable. Powell left the room for ten minutes to confer with Attorney General John Beverley Robinson.

Returning, he handed the sheriff the writ with a note on the back saying that the Act under which the commitment had been made allowed no bail or mainprize. Gourlay must remain in custody till the August assizes, for there was only one clearance of the jails per year. Anyone could see that the 80-mile trip in winter weather had been pointless in the first place. Powell might just as well have given that pronouncement when Gourlay's lawyer applied for a writ. Gourlay said that it was all part of a plot to break his spirit and make him leave of his own accord.[27]

The Ancaster *Upper Canada Phoenix*, of which only a few copies have been preserved, raged: "Mr. Gourlay is imprisoned illegally! illegally! illegally! Lieutenant-Governor Maitland can render himself extremely popular by interposing on Mr. Gourlay's behalf. As a chief Magistrate he ought to be vigilant, and see that he is not misled by the shallow artifice of violent men in power. The public feel a strong disposition to place an unlimited confidence in the Administration of Sir Peregrine Maitland. May he not abuse this confidence by taking part with those who being under the guidance of violent passions, set the law and all decency at defiance."[28] Maitland did nothing. He was no maverick from the circle that had bred him.

With no hope of redress in Upper Canada, Gourlay wrote to Montreal, Edinburgh and London for the best legal advice obtainable on the point that the Canadian Seditious Alien Act did not apply to a British subject. Sir Arthur Pigott declared that not only should the chief justice of Upper Canada have granted him his liberty when he applied for habeas corpus, but that Gourlay had grounds for a good action against the men who had imprisoned him.[29]

On January 14 the magistrates met in the courthouse for the Quarter Session of the District of Niagara.[30] Thomas Clark urged them to present a loyal address to Sir Peregrine, indicating that they repudiated any former support they had given Gourlay. The echoes of his passion reached almost to Gourlay's cell. Under Clark's tutelage, twenty-six magistrates and

---

[27]*The Banished Briton*, #16, p. 167; W. R. Riddell, *The Life of William Dummer Powell*, p. 118.

[28]The *Upper Canada Phoenix* rose from the ashes of Joseph Willcocks' *Upper Canada Guardian* when Richard Hatt bought its press. Richard Cockerell was editor.

[29]*The Banished Briton*, #16, p. 167; *Statistical Account I*, pp. x, xxxiii, cccx, cccxvlll. Pigott was attorney general in the brief Ministry of All Talents, 1806.

[30]Various powers that in New England were vested in elected town meetings were in Upper Canada delegated to the Courts of Quarter Sessions. Magistrates were men with sufficient income to render them independent of the hardships of the average settler and in the end they became an appendage of the Family Compact. F. Landon, *Western Ontario and the American Frontier* (Ryerson, 1941), p. 233.

twenty-eight others put their names to an address declaring that they were happy with the existing constitution and that the Niagara District was not the centre of folly and sedition. Eight of them, including William Dickson, in one of the inexplicable actions of history, refused to sign. On January 26 fifty-two residents of Queenston including Dr. Cyrus Sumner and Peter Hamilton drafted a similar address.[31] Robert Gourlay seemed deserted by all his staunchest supporters.

Anyone may change his mind when convinced it is for the public good. These men believed Gourlay was a threat to peace. Because he had not hidden the fact that he admired Cobbett, because he said he had seen Hunt shamefully abused at a Wiltshire country meeting a few weeks before he came to Canada, because he supported parliamentary reform, they said he was revolutionary.[32] Because he had been in Ireland for three days during the rebellion, they said he was a rebel. He denied any subversive intentions. He wrote another letter lashing out against his detractors.[33] William Dickson, who now stood in judgement against him, had recently published a pamphlet that his brother Thomas said had been written by Dr. Strachan and a respectable tradesman had been ruined by it. Thomas Dickson, who had supported his first writings, had changed because he might lose a profitable government position. William Hepburn, who had sworn that Clark had paid his dollar to send a petitioner to England, who had told Gourlay he was right to suffer imprisonment rather than flee, who had said that Gourlay should be allowed a dollar a day expenses by public subscription while he was in jail, William Hepburn had carried the loyal Address of the Queenston Residents to York to present to Sir Peregrine. As a reward he was asked to stay for dinner and sit beside the Lady Sarah. Under Smith's administration these men had been ready enough to support Gourlay but now they turned against him for fear of the new governor's displeasure. Was that sufficient excuse for jailing another to save themselves? On March 16, 1819, Gourlay wrote from Niagara jail to Thomas Dickson and William Hepburn: "Every man ought to consider that the safety of one cannot be honourably purchased at another's expense".

Some of those who had deserted Gourlay now retracted publicly. Dr. John Lefferty, writing from Stamford for publication in Ferguson's *Spectator*, repented "in the bitter anguish of my heart" that he had signed the Queenston Address to His Excellency. "As the apostle Peter wept in bitterness, for having thrice denied his Lord and Saviour, and received pardon, and was taken into favour,—I likewise hope for your pardon, and to be received into your former favours and confidence;". Others went to Gourlay

31*The Banished Briton*, #26, p. 340.
32*Statistical Account I*, p. ccxii.
33*The Banished Briton*, #26, p. 342.

in prison to ask forgiveness for their desertion. On April 6 he wrote again from prison for publication in the *Spectator*, "To the Worthy Inhabitants of Niagara District", saying that because of the assurances he had received he was certain that the people were no longer deceived by the magistrates about the sincerity of his motives.[34]

In April the grand jury was scheduled to meet in quarter session and to inspect the jail. Gourlay prepared a brief to present to them asking that an exercise yard be provided for the prisoners according to John Howard's recommendations forty years before. The grand jury did not come near his cell, for the chairman was the brother of one of the men who had committed him.[35]

By now he had been in jail for four months. The long confinement was beginning to have its effect on him, though his accommodation as a political prisoner was commodious compared to that of the others. The broom-making that occupied the other prisoners was work ill suited to his temperament. He commented dryly that he was more adept at making the kind of broom that would sweep the country clean. Since his youth, he had warded off his tendency to nervous debilitation by huge amounts of fresh air and exercise. The health that had returned to his large frame after his illness in Clark's home in July 1817 began to leave him. He resorted to medicine for relief but that was not sufficient.[36] His friends, seeing the change that was coming over him, received permission from the sheriff for him to walk in the corridor and sit sometimes in the fresh air at the door of the jail in the evening when no one was about. There, while the mourning doves whooed softly in the cedars, he enjoyed what he could.

"The summer evenings in Canada are celestial. While I had yet free range of the prison, it was my custom to sit from seven till ten at night, in the door-way, noting the course of nature, and inhaling the very air of heaven, balmy, and sweet, and invigorating. As the sun went down, a succession of tints, fine beyond what our [Britain's] grosser atmosphere can ever exhibit, insensibly stole off with the light of day. The whipperwill, bird of the evening, then arrested attention, booming through the sky with impetuous sweep. By the first starlight he rested on some stump, or fence, or fallen tree, to mark the watches of the night, with the lively repetitions of his name. Brilliant corruscation of electric fluid now flashed across the starry firmament:

[34]*Ibid.*, p. 346.
[35]*Statistical Account II*, p. 398. Likely Thomas Dickson. After the battle of Chippawa, July 5, 1814, Thomas Dickson told Major Secord that the Americans were welcome to the country. He went to Montreal till after the peace. *The Banished Briton*, #26, p. 342.
[36]*Ibid.*, #25, p. 325. Part of the trouble seems to have been that age-old scourge, constipation, that clogs the blood and angers the nerves.

sublimity took possession of the mind: every worldly care was forgotten: and it was more than contentment—it was—peaceful joy, to
sink at rest even within the walls of the gaol of Niagara."[37]

Thus he consoled himself with his love of nature.

To fill in the time, he sent for the manuscript of Barnabas Bidwell's
Sketches of Upper Canada that Bidwell evidently could not get published,
and had offered to Gourlay for use in his statistical account of the province
that had imprisoned him. It did not come in time for him to work at it.
He dreamed of what he would do when this exasperating period of inactivity was over. The first thing would be to consign the statute-book of
the province to flames, for the law which imprisoned him was not the only
one that needed repeal. He would take the law out of the hands of the
lawyers.[38] He also wrote more mud-slinging letters to the *Niagara Spectator*. Though he was confined to jail, he was not held incommunicado and
in Canada received the same privileges as the British reformers Burdett,
Cobbett and Hunt. On June 1 he answered a letter from James Durand in
which he asked "lickspittle" Durand why he was no longer calling for the
elimination of abuses. Why had Durand voted the Gagging Bill into being?
Did Mr. Durand know that in Britain 400,000 signatures had been collected
against the British bill and only 30,000 in its favour? Let Mr. Durand, the
Gagging Bill apologist, look to that! He bade Durand to be off to "Sir
Peregrine Maitland! Off, off, you Spartan dog—off to Little York, and
make clean the parliamentary journals, by gobbling up your own vomit."[39]
That letter would have been vivid language even for Cobbett who was
preparing to return to an England that his friends said was safe for him
now. He took with him the bones of Thomas Paine for burial in the land
of his birth, but the English authorities refused to allow them to be in-
terred.

April passed and May was nearly over. The government was still short
of money. On June 7, Sir Peregrine announced that he would call parliament. This evoked a fourth *Address To the Resident Landowners of
Upper Canada*, May 20, 1819. It was longer than the others for there was
plenty of time in jail for writing.

"Gentlemen, I have thrice before now addressed you: always in sincerity, and never without considerable effect. My first Address was so

[37]*Ibid.*, p. 401.
[38]This contempt for law was common to protesters like Voltaire, Godwin, and the "noble
savage" school of thought that deplored the evils of over-government and praised the
natural instincts of mankind. Twelve innocents of nature sitting on a stump, scoffed
Gourlay, could strike out of the law books such monstrosities as trial by battel and
selling wives in a halter. *Ibid.*, cccxxxvii.
[39]*The Banished Briton*, #29, p. 401.

. . . palpably beneficial to the province, that it must remain to excite surprise, how a single voice could be raised in opposition. My second Address . . . was to advise parliamentary inquiry into the state of the province, and the appointment of a commission to carry home the result. My Third Address, holding the same object in view, sprung from a momentary impulse, occasioned by the sudden and extraordinary breaking up of parliament. Wherever the light of information reached, this Address was electric, and thousands of the most loyal hastened to conform to its *dicta*. Horrible to relate, there were found among your own representatives, men, who . . . bereaved their constituents of the most valuable privilege—that of employing rational means for carrying a petition to the foot of the imperial throne.

"Gentlemen! by the acts of your own representatives, you can no longer boast of freedom: you are, in fact, this moment slaves. Alas! am I not myself a striking witness of this truth?—a prisoner, locked up at the capricious mandate of my personal enemies; as which of you may not be. And is there now an honest and independent man among you who would raise his voice against inquiry, who would prefer darkness to light, who would sit in ignominious silence while such things are? Was it to be enslaved that you came from the United States of America, from Britain, and from Germany? Was it to be enslaved, that you here supplanted the native Indians, noble and free? Was it to sow the seeds of despotism that you lifted the axe to clear away these woods? Was it for the growth of tyranny and oppression, that you let in the solar rays to warm and fertilize the turning earth? . . ."[40]

Again he asserted his loyalty and his conviction that a commission of inquiry to go home to Britain to seek redress of grievances would not break but enhance the connection between Britain and her colony.

He had more unsolicited advice for the parliament to be convened on July 7. On June 10 the columns of the *Niagara Spectator* were filled with an "Address to the Parliamentary Representatives of the People of Upper Canada".[41] He urged them to tax wild land as well as occupied. He said that the Canadian parliamentarians, particularly those members who held thousands of acres, were following William Pitt's practice of taxing the middle and poorer classes and letting the rich go free. He said that he had given much thought to taxation systems. Land should be the main basis of revenue, for commerce was not sufficiently advanced to produce much. He begged the parliamentary representatives to adopt a three-year plan for bringing in enough men to build the necessary St. Lawrence canals. The first year's budget would be £50,000 and the total expenditure up to

40*Statistical Account I*, cccclxvii ff.
41*Ibid.*, ccclxxvii ff.

December 1, 1822, would be £800,000. This he proposed to an assembly that could not find £8,271 for the still unapproved civil list.

Just as this address was published, a friend came from York with a copy of Sir Peregrine's Speech from the Throne. Sir Peregrine referred to the fact that no change had taken place in His Majesty's indisposition. He expressed his regret that Queen Charlotte's death had brought to a close a long and industrious life. He regretted that the royal assent to the bill recommending the establishment of a provincial bank had not arrived in time for legal promulgation and asked for a further enactment to render it available. He suggested that the government consider assessment of land now exempt, and he recommended a change in the Road Law. He announced that His Royal Highness, the Prince Regent, had authorized the governors of both Canadas to bestow 200 acres of land on members of the provincial navy and militia who had served during the late war, but that he was withholding this bounty from any of the militia who had attended the late Convention of Delegates, a proceeding that he considered was subject to severe animadversion.[42] The legislative council humbly thanked Sir Peregrine for his pronouncements. After the legislative assembly deliberated, it ventured to "lament" that any of His Majesty's subjects should forfeit their claims upon the bounty of their government because of their attendance at the Convention of the Friends of Enquiry, but that was all.

Immediately, Gourlay set pen to paper to fill columns of two successive issues of the *Spectator*, June 17 and 24, with a second "Address to the Parliamentary Representatives of Upper Canada". Again he outlined his views on taxation. He said that England might have made the change from war to peace with ease instead of hardship. During the war England had wasted millions of pounds and hours of manpower just to restore the Bourbons to France and Ferdinand to Spain. To find money the government had adopted a funding system to generate fictitious capital. When the time came to put Louis XVIII on the throne, a British banker had advanced the money. Why was money suddenly scarce? The British landed interest—the Duke of Devilry, my Lord Lubber, and Sir John Sinecure— kept rents up by raising the duty on grain. There would be money if the lords, knights, bankers, fundholders, and the rest of those at the top would decrease expenditures from £30,000 to £20,000 yearly, but "The further down, the greater is the misery; but the further down, the less is that political influence, which can guard against approaching ruin, and the more removed are individuals from the sympathy of those who have it to wield." He begged the parliamentary representatives to realize that their province did not exist for itself alone. He begged them to think of the fever-ravaged poor of Ireland whose desperate plight was being perfunc-

[42]June 7, 1819, *Jls. Leg. Ass.*, O.A. Report #10, 1913, p. 99; *Statistical Account I*, cccxcvii ff.

torily considered by the British House of Commons and to whose beggary
he had been calling attention for the last five years.[43] He begged them
to consider that "The first question in political economy should be, can the
mass of the people live comfortably under this or that arrangement? But
this most necessary question was forgotten, and many of the people have
perished."[44]

Political economy was a term so new even in academic circles that it had
echoed strangely from the lips of Dugald Stewart in the classical halls of
Edinburgh only a few years before. Political economy itself is formulated
according to the prejudices of the writer. Malthus and Ricardo were
political pessimists who assumed the inevitability of mass privation and
great inequality. Gourlay was a political optimist who thought that the
economy could serve the interests of the mass of the people at the bottom
of the pyramid of society. Of the government leaders who would read his
words Sir Peregrine was a Tory militarist; William Powell was a legal man;
John Strachan was "a dry morality preacher" in religion but no saint in
politics; canny William Dickson feared to endanger the business prospects
that were making him the richest man in the province; careful Thomas
Clark disliked visionary enthusiasts; James Durand was a reformed reformer; Philip VanKoughnett's consuming passion was hatred of Yankees;
Jonas Jones raved against the imbecility of the masses. All these men were
upright men, good for governing on a provincial level, but Gourlay was
asking them to consider the age-old problem of poverty, to re-think their
responsibilities in life.

He published another letter on June 28. "Men! should I call them men,
whose conduct has no parallel, if we descend not to the brute creation, but
in that of wilful pettish children, . . . Poor creatures! poor Peregrine! what
a poor pickle they'll make of it at last!" There was talk about renewed war
with the States over Jackson's arrest of British Arbuthnot and Ambristier
in Spanish territory in his war against the Seminoles. If the Americans
were to invade Canada again, now would be the time to invade a country so
spineless that it would not adopt plans to give employment to millions of
deprived souls.[45]

As soon as the second "Address to the Parliamentary Representatives"
was published, the legislative assembly, led by James Durand and Jonas
Jones, passed a resolution calling it "a most scandalous, malicious and
traitorous libel, tending to disturb the peace of the Province and excite insurrection against the Government."[46] Sir Peregrine ordered Attorney
General Robinson to arrest Bartimus Ferguson again. After Ferguson's

43For official attitudes to the plight of the Irish see *The Great Hunger* by Cecil Woodham-Smith (Hamish Hamilton, London), 1962.
44*Statistical Account I*, ccccvii.
45*The Banished Briton*, #34, pp. 470-1.
46July 5, 1819, *Jls. Leg. Ass.*, O.A. Report #10, 1913, p. 173.

previous arrest he had been told that he would not be prosecuted for any publication as long as he kept the signed manuscript to prove that he had not written the article himself. When the second "Address to the Parliamentary Representatives" was published in his paper, Ferguson was a hundred miles away on the Talbot Road.[47] On the night of July 13, 1819, he was dragged out of bed and brought back to Niagara jail. Three days later he was taken to York before the three judges. Because he was a stranger in York, he could not raise bail there. The judges gave him permission to file a plea for bail in the Niagara district, and after screwing his courage for the long hot trip back by a few days of disporting himself in York inns, sheriff Merritt set out on the return journey with his prisoner.

On the way back Merritt ill-treated Ferguson as he had not done on the journey to York. Ferguson attributed this to the dressing-down Merritt had received from Attorney General Robinson for allowing Gourlay pen and paper in jail so that he could spirit out letters for publication. Merritt grumbled that he could not stay at the jail every minute watching Gourlay for he had other duties as well. Robinson threatened Merritt with the loss of his job and it would have been no trouble at all to find another reason for dismissing him. In addition to his salary as sheriff Merritt was collecting money as a half-pay officer, though to get it he had to swear that he held no office under the government. Like bidding up Lord Morton's bull at an auction, this was a practice condoned among friends.

Now Swayze said in the assembly that a company of Gourlay's friends had promised to indemnify Ferguson for all costs and damages that he might incur because of his boldness in publishing Gourlay's letters. Editor Ferguson in the columns of his paper immediately gave the lie to Swayze. "This declaration," wrote Ferguson, "was false and infamous—infamous because it was false. I stand alone in these prosecutions, and am willing to submit to a jury whether a free press shall be maintained in the District of Niagara; if they decide not, the penalty falls upon me alone."[48] The previous December 1 executive councillor Strachan had written to Dr. Brown, his former mentor at St. Andrews University: "A Character like Mr. Gourlay in a quiet Colony like this where there is little or no spirit of inquiry & very little knowledge may do much harm & notwithstanding the check he has receved (sic) he has done harm by exciting uneasiness irritations & exciting unreasonable hopes."[49] The Province no longer lacked the spirit of inquiry that is the salvation of the human race, and some were even willing to suffer for the right to inquire.

Ferguson had retained a Niagara lawyer, Mr. Taylor, to defend him.

[47]*The Banished Briton* #34, p. 474; #7, p. 68; *Statistical Account I*, p. xl.
[48]*The Banished Briton* #34, p. 476, taken from *Niagara Spectator*, July 29, 1819.
[49]Strachan to Dr. Brown, York, Dec. 1, 1818, G. Spragge, ed., *The John Strachan Letter Book*, p. 185.

Gourlay urged him to traverse and employ the argument used by Lord Erskine in the case of Cuthell.[50] Gourlay himself intended for the third time to conduct his own defence. Both Ferguson's trial and his were now only a month away.

[50]*Statistical Account I*, p. xiii.

20

The Banished Briton

The Gagging Act had terminated the township meetings and now the jail regulations were tightened to prevent Gourlay from releasing any more scripts for publication. Six weeks before his trial was scheduled, he was no longer permitted to walk in the corridor of the jail or sit at the door in the evenings. The only communication he was allowed with his friends was through the grille in the cell door. The July heat had begun. By mid-day the jail was like a furnace and a sodden silence hung over it. In the spring the prisoners had sung, sometimes called to each other. Now any voice that attempted a tune soon faded away. If Gourlay pressed his face to the aperture in the thick door that opened to the corridor in search of a movement of air, a stench drove him backwards.[1] He began to lose weight. He began to lose will. Sometimes he could hardly remember why he had to stay in this jail.

It was the hottest summer that the Niagara peninsula had ever known. For six weeks no rain fell. Cicadas screamed and shrilled in every tree. So many grasshoppers flew in the grass that whole fields seemed in motion. The dog days were on the province, the days when the muggy southern air moved north, using the Mississippi valley like a horizontal funnel. The thermometer rose to ninety-five. In the jail that had been built on swampy ground just outside Niagara, Robert Gourlay lay panting for air, too weak to write. The sultry air rested on his chest so heavily that he breathed with difficulty. Exertion was impossible without a bath of perspiration. No one had devised this torture for him; it had just happened. "One morning, while gasping for breath, I besought the gaoler to let me have more air, by throwing up the window. 'You are no gentleman,' he said; 'you gave that letter out of the window, and I will come presently to nail it down.' Happily a friend soon after called upon me and through his interference the window was put up."[2] The letter was the one he had written to protest Ferguson's prosecution for publishing his writings.

[1]*Statistical Account II*, p. 307.
[2]*Ibid.*, p. 393. Page 181 quotes the Duc de la Rochefoucauld's opinion: "It is less the intensity of the heat, than its peculiar nature, which renders it altogether intolerable."

He roused himself to write a letter protesting conditions in the jail, but the *Spectator* could not print it for its editor had sworn not to publish any more of Gourlay's writings. Gourlay's opponents had the run of the columns now. Uneasy at the obvious discrimination, the *Spectator* published its misgivings at inserting only writings so obviously weighted on one side.[3]

As the burning days wore on, Gourlay pulled some paper in front of him and worked to prepare his defence for his trial on August 20. To prepare it, he had to know the charge against him, and his opponents were maintaining as much secrecy now as they had before his former trials. When his friends came to the door, he asked them if they had found out yet. If it were sedition, he was confident that he could free himself again. In the sweltering heat, he tried to think. He took up his pen but the paper lay in front of him as blank as when he had pulled it out. His head ached intolerably. He could no longer sustain a period of reasoning. Now his friends came with the news that it was unlikely that he would be charged with sedition for twice he had vanquished the Crown on that count. He would be charged with not leaving the province when ordered to by William Dickson. He could not believe his ears. After they left, he tried to think of a defence against such an absurdity. Not leaving the province when ordered by Willie Dickson, on the oath of a man as insignificant as he considered Isaac Swayze! The thought was so ludicrous that he burst into a fit of wild, uncontrolled, convulsive laughter that echoed into the corridor and back to him again.[4] Who ever heard of trying an intelligent human being on such a ridiculous basis?

Then mercifully, about three days before his trial, the weather changed. The rain came, the temperature dropped, human beings breathed deeply again. The fresh air revived the prisoner in his cell. He stirred about. He tried to work on a defence against this picayune accusation but it was impossible. He thrust the papers into his pocket and lay down again. After the brief respite, the heat returned as oppressive as ever.[5]

On August 19, Bartimus Ferguson was tried. A special jury found him guilty of publishing sedition and he was returned to jail to await pronouncement of his sentence.[6] It was obvious that the sentence meted to the editor would be determined by the outcome of Robert Gourlay's trial. Ferguson had a home and family in the district and because of them he might be expected to forget his crusade for the liberty of the press.

On Friday, August 20, 1819, the door of Robert Gourlay's cell was opened and he shambled out of its close atmosphere. He was outside his

---

3*Niagara Spectator*, August 19, 1819; *The Banished Briton*, #34, p. 477.
4*Statistical Account I*, p. xv.
5Gourlay's torture because of the heat was of course fortuitous. John Thelwell, lecturer and orator of the London Corresponding Society, was lodged before his trial in 1794 in the charnel house used for prisoners who had died of cholera.
6*The Banished Briton*, #34, p. 477.

four walls at last, even if it were only in the corridor. Into the crowded courtroom he went. Every window had been opened to alleviate the heat for those who had come from all over Upper Canada and New York state to see the great Robert Gourlay who had defended himself successfully twice. Part of the crowd was hoping that he would confound officialdom for the third time and part was hoping that this enemy of the country would receive his due. Everyone knew that the occasion was crucial except the man who was on trial.

The man who had been confined in one room through six weeks of cruel and humid heat found the air in the crowded courtroom more intoxicating than alcohol.[7] His brain became light and airy and instead of being able to concentrate on his defense, all he could do was breathe in great gulps of air and feast his thoughts on the prospects of immediate freedom. He could not concentrate on his . . . on his . . . what was the word he wanted? What did he want to do? Why was he here?

He could not quite recall the reason for his being here, for the lightness in his brain would not allow it. Instead his thoughts roamed back to the days of his youth. He was running, laughing and happy, under the trees that surrounded the little school at Craigrothie. He was walking through the halls of St. Andrews with his friend John Dean looking up at the portrait of wicked Cardinal Beaton. He was urging the drunken typesetter in Bath to hurry so that parliament would be moved to amend the poor laws. He was standing in the doorway of Deptford Farm watching the lark springing from the field and the little daisies in the lawn before the house that sheltered his wife and children. Again he tried to think of the word for which his brain was searching and again he failed.[8]

Now the jury was being sworn in. Knowing on how many occasions in Britain juries for political trials were packed shamelessly, Gourlay had taken precautions against such an eventuality.[9] His friends had handed him lists of those inimical to him and he had people willing to swear that these men were prejudiced against him and therefore disqualified for jury duty. Now he was unable to challenge them because the fresh air was causing delirium in his brain. All his determination to watch every move at this trial was gone. It was customary during trials to draw jurymen in

---

[7]*Ibid.*, #25, p. 325; *Statistical Account I*, p. xv; *Appeal*, p. xix.

[8]Much of this description has perforce been taken from the first chapter of John Dent's *The Story of the Upper Canadian Rebellion*, 1885. Dent has long been a thorn in the side of Canadian historians for he gave few sources for the extraordinary bits of history he preserved, such as this account of Gourlay's trial. The thoughts that went through Gourlay's mind are recorded nowhere else. When Dent described the trial, two people who had witnessed it were still living.

The following year Dent's defence of Gourlay was attacked by John King, *The Other Side of the Story* (Toronto, 1886).

[9]*Statistical Account II*, p. 342; Dent, *The Story of the Upper Canadian Rebellion*, I, p. 31; G. D. H. Cole, *The Life of William Cobbett*, p. 152.

rotation from the various townships in the district. The man who was his own defense lawyer could not gather his wits to declare that this time they had not been selected from the usual square of territory but from a straight line, twenty miles long, populated by many of his political opponents. His friends, watching anxiously for him to challenge the choices, saw that the sheriff drew the names from a pile of folded slips in the glass where a little heap seemed separate from the others. Gourlay challenged no one, not even John C. Ball who had called at his cell just before the trial to accuse him of being the worst kind of culprit, worthy of the worst kind of punishment which could be devised.[10]

The jurors chosen were William Pew, John Grier, William Servos, James B. Jones, Ralfe M. Long, David Bastedo, John C. Ball, John Milton, James Lundy, William Powers, Peter M. Ball and John Holmes. The man who should have challenged this list was blinking with watery eyes at the crowd that had gathered for what they expected would be a show as brilliant as those in Kingston and Cornwall. Gourlay could not remember that he was a great leader in the fight for free speech, for a vote for every man who could read, for education for those who could not, for food for empty bellies, for the right of the people to watch what their elected representatives were doing, for liberty against tyranny.

"How say you, Robert Gourlay, are you guilty or not guilty?" He started in his seat and for a moment stopped the uncontrollable trembling with which he had been seized. The word he could not recall came to him now. The word was "protest". He was here to protest against a charge so ridiculous that it was beneath his intelligence to consider. He must protest against the charge, but all he could make was a meaningless mutter that was interpreted as "Not guilty."

"Are you ready for your trial?" Every man in that crowded courtroom, the judge included, could see that the man in the prisoner's dock was in no condition to stand trial. Where was the vital Robert Gourlay who had confounded his accusers at two previous trials? "The personal appearance of the prisoner had undergone a woeful change during his confinement. Had his own wife seen him at that moment it is doubtful whether she would have recognized her lord. Could it be possible that that frail, tottering, wasted form, and that blanched, sunken-eyed, imbecile-looking countenance were all that was left of the once formidable Robert Gourlay? The sight was one which might have moved his bitterest enemy to tears. His clothing, a world too wide for so shrunken a tenant, hung sloppy and slovenly about him, and it was remarked by a spectator that he had aged fully ten years during the six months that had elapsed since his journey to York in the previous February. His limbs seemed too weak to support him where he stood, and as he leaned with his hands upon the rail in front of

[10]*Statistical Account I*, p. xv; *Statistical Account II*, p. 342.

him his fingers twitched nervously, while his whole frame visibly trembled. The saddest change of all had been wrought in his once fine eyes. They were of light grey, and their ordinary expression had been more sharp and piercing than is commonly found in eyes of that colour. They had been clear and keen, and expressive of an active and vigorous brain behind them. At present they were wandering, weak and watery, altogether lacking in lustre or expression. They told their sad tale with piteous brevity. The brain was active and vigorous no longer, or, if still active, was so to no definite purpose. The spark of reason was for the time quenched within him. His oratory and his writings were no longer to be dreaded. The man whose large presence had once carried about it unmistakable evidences of physical and mental power had been reduced to a physical and mental wreck. No man in that closely-packed courtroom was now more harmless than he."[11]

What was the charge against the prisoner who had been trying for six months to discover what it was so that he could prepare his defense? The charge was that Robert Gourlay had not left the province when ordered by councillors William Dickson and William Claus, on the oath of Isaac Swayze that Robert Gourlay was a seditious alien because he had not been an inhabitant of the country for six months preceding the order.

Who was the judge in charge of the trial of a man who was obviously ill? Sitting in the chair behind the big table on the daïs, and not feeling too well himself in the heat of the crowded court,[12] was Chief Justice William Dummer Powell, who had congratulated Gourlay a short time before on his project for a statistical account of Upper Canada and who had opposed Strachan when he labelled the project mischievous. Powell was a sharp man in a land deal and a good lawyer who liked to beat the Niagara faction and keep ahead of the rector of York. In any other situation except his present position in an emerging country, William Powell would have been a minor administrator whose tendency was to hold liberal sympathies. He had refused to declare that Gourlay's meetings violated the law. He had not liked the Seditious Alien Act when it was passed in 1804, but he had not said no when he was asked to write with his own hand the reinterpretation that had sent to prison for nine months without bail the man in the prisoner's dock. A true bill had been found against the prisoner on the charge that he had refused to leave the province when ordered and he

---

[11]J. M. Dent, *The Story of the Upper Canadian Rebellion*, I, p. 31.

[12]Mrs. Powell to G. W. Murray, York, Aug. 22, 1819, *Letters of Mrs. W. D. Powell*, 1807-1821, *Niagara Historical Society*, #14: "I am anxious to know how he supported the heat of a crowded court (the summer intensely hot & Mr. Powell unwell,) when the thermometer was 92 in the shade. The trial of Gourlay for sedition has I understand collected a vast concourse of people as well from the state of New York as from different parts of the province." Powell himself had been arrested soon after his arrival in Canada in 1779 for treasonable correspondence with the Americans. *Statistical Account II*, p. 517.

had not disallowed it. Even a judge finds it difficult not to lean to the right or the left as he treads the tight-rope of justice, and Powell seems to have been pushed to the right by those of his peers who were convinced that the country was in danger.

Who was prosecuting this trembling man who was obviously in no fit state to stand trial? It was not Henry John Boulton who had twice failed to vanquish the defendant. The prosecutor was Attorney General John Beverley Robinson who five years before, at the age of twenty-three, had condemned eight men to death at the Bloody Assize of Ancaster. The Scottish Treason Trials had condemned men only to deportation; the English Treason Trials had acquitted the men when the judge asked for the death penalty for them; the Canadian Treason Trials, under John Beverley Robinson, sent eight men to their death. The handsome suave attorney general had just returned, all book and rules, from finishing his legal training in England. It was he who had been chosen to save his country from the man who was disturbing its tranquillity. His mind was keen. It had a tremendous faculty for penetrating circuitous legal terminology, but it was not given to cosmic contemplation of changes that would build a new society or correct existing evils. Already he evinced the authoritative strength native to a man who had formed his opinions early and would never see any reason to change them. He had had ample evidence of the outcome of trials when the accused was in full possession of his strength. Gourlay's Kingston Address to the Jury had been hailed as a seldom equalled "specimen of bold, energetic composition".[13] Robinson had prepared this case well even if the issue was simple. The jury had simply to declare that this man had not left the country when ordered, and there the man sat. He was anxious to perform, for already he had his eyes on the position of elderly Chief Justice Powell who did not sit so firmly in the favour of Sir Peregrine Maitland as he had with the previous lieutenant governor, Francis Gore. He could not help seeing that the prisoner, the man who had the temerity to conduct his own defence for the third time, was unable to give much opposition.

Who was Gourlay's accuser? It was Isaac Swayze, ex-Quaker; Isaac Swayze, determined to rise in society from the troubled depths into which it had been his lot to be born; Isaac Swayze, who seemed fated to live his life scrambling up the sides of the gorge of existence; Isaac Swayze, fanatical anti-democrat, useful for carrying out some little business like spying on meetings; Isaac Swayze, Dickson's cat's paw. In his two previous trials Gourlay had at least been honoured to have the King as his opponent. This time Gourlay's accuser was Isaac Swayze who had perjured himself to please the official circles he loved. He knew, as well as the others, that Robert Gourlay had lived in the province of Upper Canada for more than six

---
[13]*Pittsfield Sun*, Sept. 9, 1819.

months, but he did not consider that what he had sworn was perjury. It was patriotism.

The great fighter hardly heard Robinson examining the two main witnesses for the Crown: William Claus, whose own actions in land deals in the Indian Department were as questionable as anything Gourlay had ever done; and William Dickson, Machiavellian enough to find a man willing to commit perjury and therefore guilty of the same offence himself. Dickson had been the first to urge Gourlay on his course of criticism. Now that he had secured the admission of Americans, he need no longer be a critic of government. Furthermore, if he continued to support Gourlay, he might himself be accused of complicity. He would be obliged to resign from the legislative council, the Upper Canadian equivalent of the House of Lords.

What would Robert Gourlay's defence be? In his pocket he had the draft of a speech that he had prepared in the heat in the expectation of being tried on a charge of sedition. When he had heard that the charge was refusal to leave the country, his mind had not been able to adjust itself to a new line of thought. He had lapsed into a mist of happy boyhood memories. Now he tried again to force himself to remember where he was and why he was here. He rose to argue his defence. As he struggled for coherent expression, an immense contempt for his accusers filled his mind. How could he marshall his thoughts to argue against—surely the accusation must be sedition; it couldn't be refusal to leave the country when ordered by Willie Dickson. He must protest a charge so absurd that it was a hideous mockery of justice. For a moment he could think, and now he realized the grandiose joke that fate had played on him who had started out twenty years before with a grand dream of righting the wrongs of the British poor. His twitching nerves could not sustain the thought of his ludicrous situation. He burst into a peal of the same maniacal laughter that had escaped from him in his cell a few days before. It was a cachinnation that chilled the cheeks of the listeners.[14]

The outburst relieved him temporarily. He attempted to argue his case. He tried to introduce as evidence a copy of the *Salisbury Journal* of April 1, 1816, in which he had expressed the difference of his political opinions from those of Cobbett and Hunt, whose friend his accusers said he was, and his disapproval of large irregular meetings like those of Spafields, whose violence they said he had fomented.[15] His evidence was dis-

---

[14]There seem to have been two outbursts, one recorded by Gourlay, *Statistical Account I*, xv, as a "fit of convulsive laughter" in jail when he was told that he would likely be tried for not leaving the province when ordered, and the one noted by Dent, *The Story of the Upper Canadian Rebellion*, I, p. 37, as a "senseless bray" during his trial. Gourlay later attributed his "imbecility in court" to the "joint effect of disgust and disease". *Statistical Account*, II, p. 498.
[15]*Statistical Account I*, p. ccclvi.

allowed by Attorney General Robinson. He tried to say that the sheriff had refused to permit him to publish in the Upper Canadian papers the opinion of the illegality of his imprisonment as declared by Sir Arthur Pigott. The chief justice informed him that he could be permitted to speak only on questions of law. Those had all been decided before, and the accused could find no mental or physical strength to argue against these restrictions. He sat down with his disordered papers before him. His defense was reported in the censored *Niagara Spectator* as "idle and absurd to the extreme and such as disappointed the expectations of those who still adhered to the man".[16]

The verdict of the jury was "Guilty". What else could a jury do, packed or not, when asked to decide whether a man who had openly refused to leave the country had or had not left the country when ordered? This is what it could have done. It could have heeded Gourlay's directive in his letter to the *Niagara Spectator*, March 23, 1818, citing the case of William Penn.[17] A judicial snare had been laid for Penn, but the jury refused to declare him guilty. The judge tried to starve that jury into submission, but still it refused to declare Penn guilty. Penn went free. This jury of twelve Upper Canadians in 1819 declared that Robert Gourlay was guilty.

Before he pronounced sentence, Chief Justice Powell delivered a lecture on the even-handedness of Justice which metes impartially to rich and poor.[18] In Canada, he said, no man could be allowed to run counter to the law. He told Robert Gourlay to turn his great abilities to some practical account in order to win distinction and happiness for himself. Then he pronounced sentence on the man who had been born with more than his fair share of the thistle in his blood and a martyr's unconcern for his own well-being. Robert Gourlay was banished from the province of Upper Canada on pain of death if he returned, banished, not even in a Moses-anger to be admired, but by a legal quibble.[19]

Gourlay made one more effort to speak for himself. He told the court that his letters had been suppressed by the sheriff but that his enemies had been allowed to fill recent newspapers in order to prejudice the public mind against him just before trial. The chief justice informed him coldly that he could prosecute the sheriff.

Gourlay returned to the prisoner's dock. He had understood the verdict that he was guilty but he was still confused about the charge. Lean-

---

[16]*Kingston Chronicle*, Sept. 3, 1819, copied from non-extant issue of the *Niagara Spectator*. The trial was overshadowed by black-bordered accounts of the death of the Duke of Richmond from rabies resulting from the bite of a pet fox.

[17]*The Banished Briton*, #21, p. 229.

[18]J. M. Dent, *The Story of the Upper Canadian Rebellion*, I, p. 38.

[19]It is likely that Gourlay's opponents considered him lucky not to be hanged. In November 1797, the execution of a Mr. McLean "effectually checked every seditious symptom among the Lower Canadians for the present". C. E. Cartwright, *Life and Letters of the Honourable Richard Cartwright*, p. 75.

ing over the side of the box, he asked a startled juryman if he had been convicted of sedition for he still could not believe that he could have been sentenced on an absurdity. Then he left the court in the company of his friends. The *Niagara Spectator* recorded the disappointment of the crowd that had come to see the great Robert Gourlay confound his detractors for the third time. All they had seen was a sick man.

The rumour spread that Gourlay was insane because of his outburst of maniacal laughter during his trial. Was he insane or was he laughing at the human comedy of the men who were judging him, the men who would go down in history as absurdities themselves because they allowed themselves to be trapped by unbalanced fear and rigid refusal to accept criticism from a man who looked at the present in order to guide the future. They called him a political agitator, a threat to the public peace. They allowed justice to miscarry in order to silence him. No historian has ever denied that the law was stretched in order to banish Robert Gourlay and some have declared that in addition he was banished illegally because he was unfit to stand trial.

Why should anyone consider that justice had miscarried for a man whose opinion threatened the established order? The men who condemned Gourlay refused to face the real issue. The banished Briton knew what it was: "a weak jury headed by a petty magistrate, gave that away, which the people in the United States, who have, for twenty years past, disclaimed the very notion of political libel, would sooner have lost their right arms than have parted with. No one would punish slander on private character more severely than I would: men in power should always be entitled, as well as others, to redress in civil suit; but, to make due reprobation of mal-administration in public affairs criminal, is what I shall protest against while I have breath".[20]

Who was to blame for the miscarriage of justice that caused Robert Gourlay to be banished? When he reconstructed what had happened, he would always end with "It all began in York", that John Strachan was "the sole beginner" but William Dickson carried it through.[21] To find where the responsibility lay, let us examine the part of each of the major figures concerned with the trial, beginning with the lieutenant governor, Sir Peregrine Maitland.

Sir Peregrine did not order Gourlay's arrest, but he was not ignorant of what was going on under his nose. It was to him on November 10, 1818, that the three judges delivered their interpretation of the word "habitancy" as used in the Seditious Alien Act. It was he who asked them to re-interpret this law and it was to him on December 16, 1818, that Isaac Swayze reported Ferguson's arrest. Maitland made no move to countermand Dickson's and

---

20*Statistical Account I*, p. xiii.
21*The Banished Briton*, #30, p. 428; #26, p. 347.

Claus's charge against Gourlay as he waited nine months in jail for the August assizes, though he had the power to do so. On July 22, 1819, he wrote to Goulburn, the permanent under-Secretary for War and the Colonies, "The fellow in a Province like this was superlatively mischievous as long as he had the only two papers that are read completely in his interest, for the Government paper, it was said out of delicacy, took no part. The liberty of the Press is however now established. . . ."[22] This part of Maitland's letter is not true, for both papers reported both sides until the *Spectator* was forced to cease printing letters favouring Gourlay a month before the trial. Maitland continued: ". . . and he was sinking into insignificance when two Legislative Councillors on the 44th of the King perplexed me by taking him up and ordering him out of the Province. He would not go and they have put him in Gaol. I am afraid this will give him a new interest for a short time, but I trust under the blessing of the Lord, that all will end peaceably. I have no sort of apprehension and intend to let the law take its course—I had intended to have pursued different measures, but that which I would not have done, it would on many accounts be imprudent to undo".[23] Maitland seemed relieved that someone else had assumed the responsibility for an action that he might conceivably be called upon to justify.

Maitland had come to Canada to perform his duties conscientiously. He found what he considered to be seeds of rebellion in the Convention and the Addresses. Sir Peregrine belonged to the class, as Gourlay noted, which was busy at the Congress of Vienna restoring crowns to heads that had done nothing to deserve them. It was not in Maitland's thinking to declare that a government had no right to exist if its actions could not bear scrutiny. Criticism and sedition were symonymous in the early nineteenth century. The term "His Majesty's loyal opposition" would not come into use till coined by Hobhouse in 1826. Maitland was not a man who was ahead of his time. Dr. A. R. M. Lower's opinion in *Colony to Nation* is that "He was crammed with the pet prejudice of religion, flag, and caste: the perfect Colonel Blimp. . . . A good deal of the leftward movement of events during the 1820's can be chalked up to Sir Peregrine's score."[24] The reformers referred to Maitland's régime as "the dark days of Maitland".[25] Nevertheless, the biographer of Sir Peregrine, Michael Quealey, argues convincingly that "Maitland was simply not as involved . . . as some polemical writers in the liberal tradition have indicated.[26]

Was it Chief Justice Powell who was responsible? He and his colleagues

22Maitland to Goulburn, York, July 22, 1819, Q 325-32, p. 92.
23Maitland to Goulburn, York, July 22, 1819, CO42/362/257.
24A. Lower, *Colony to Nation*, p. 238.
25*The Mirror*, Mar. 3, 1839.
26M. Quealey, *Sir Peregrine Maitland*, Ph.D. thesis ms., University of Toronto, p. 183.

were asked by Sir Peregrine to interpret the Seditious Alien Act of 1804. Of the three judges Mr. Justice D'Arcy Boulton was noted for severe sentences. Mr. Justice Campbell had already presided over two trials that had acquitted the man who would be prosecuted a third time. No statement of the latter two men has survived.

The third judge, Mr. Justice Powell, said that he disapproved the prosecution against Gourlay. As early as 1818, he wrote to Gore in Britain: "Your sound advice to let Gourlay run his race was in unison with all here and when I absented myself for a few weeks in the early summer I did it with the assurance that such would be the policy of the government". Just after Gourlay's committal to jail following his refusal to leave the province, Powell wrote again to Gore: ". . . the enactment subjects Earl Bathurst if he should pay a visit to this province and his looks offend Isaac Sweezy, to be ordered out of the Province by that enlightened magistrate, and upon Disobedience to be imprisoned and tried and if that disobedience which constitutes the offence is found by a Jury, to be banished under penalty of Death, should he remain or return, without the slightest enquiry into the cause or justice of the worthy magistrate's suspicion that he was a suspicious character. I almost persuade myself that the English lawyer who drew the Bill wrote in one of the clauses or for and which makes all the difference between a just enactment and the most absurd tyranny which ever disgraced a Legislative Act. . . . The knowledge of such a Law, I should think, would discourage more than fifty Gourlays."[27] Yet Powell would not declare the defect in the Act and refuse to prosecute Gourlay by means of it. Powell's only biographer, Justice William Riddell in 1924, declared: "Gourlay was brutally used, but the blame should be put where it is deserved. Powell's skirts are clear . . . all he did was, as a Judge, to interpret the Act, . . . —and no one has ever said that in the performance of this duty he acted harshly or even discourteously."[28] Not all agree. Gourlay was crushed "by as sorry a manipulation of the law and of the bench as ever disgraced a British country", says Dr. A. R. M. Lower in contradiction.[29]

The situation was a complicated one and Powell was sixty-four years of age when, as he admitted himself, "imagination & memory begin to flag".[30] He had a grudging admiration for Gourlay: "Your Mr. Hunt," he wrote to Gore after Smith's prorogation of parliament and Gourlay's Third Address, "is a fool to our Mr. Gourlay . . . ".[31] If Gourlay had implied that he had received his information on grievances from Clark and Dickson, he

---

27Powell to Gore, York, Jan. 18, 1819, Powell Papers, Toronto Public Library, Vol. 1806-1823, A 30.
28W. Riddell, *Life of William Dummer Powell*, p. 119.
29A. Lower, *Colony to Nation* (Longmans, 1946), p. 234.
30Powell to Gore, York, Aug. 5, 1818, Powell Papers, Toronto Public Library, Vol. 1806-1823, A 30.
31*Ibid.*, April 6, 1818.

continued, they in turn had used their relative to embarrass the executive government in the power struggle of which he himself was a part.

Powell was liberal enough to oppose the Gagging Bill but not sufficiently committed to forestall it.[32] He abhorred the Sedition Act under which Gourlay had been arrested: "in its present Condition [it] is the approbrium of the Colony and perhaps the only just cause of complaint existing".[33] Nevertheless he was not in sympathy with Gourlay's reforming motives: "The unhappy Man (if a Madman can be so called) . . . has been allowed from his Prison to scatter Seditious . . . libels of the most atrocious sort & firmly has drawn upon him again the Censure of the Assembly in an Address to the Governor to prosecute him & the Printer for libel—what is laughable in this is that the Govt. in Compliance having ordered the Att. to prosecute, is now condemned, by themselves for the mode adopted by that officer, by Information instead of Indictment."[34]

Powell might have resigned in protest but he had held his high position for only three years. Demotion at any age is hard to take. He had complied with the wishes of the executive government and retained his position. Nevertheless from this time, Powell's influence in government declined and the triumvirate of the right—Strachan, Robinson and Hillier, Maitland's secretary—gained an ascendancy that would last till Lord Durham was sent to investigate the unrest in Upper Canada. Whichever way Powell stepped at this time would have been the wrong way as far as his career was concerned, but in some eyes one way was more honorable than the other.

Was it then the attorney general, John Beverley Robinson, who must be censured or praised (according to the reader's point of view) for arguing his case so successfully that Gourlay was banished? Robinson had won where Boulton had failed. In the course of his illustrious career, case after successful case would pass under Robinson's gavel. Robinson was a man who loved the balance he considered an aristocracy gave to a society. He was an arrogant man filled with the consciousness of his own superiority.[35] He was and would be till the end of his days the fulcrum of an unchanging system, the epitome of the Family Compact. The man he had just proven guilty was his equal in intelligence, breeding and consciousness of class but he was a champion of popular rights. At this moment Robinson had triumphed, but the day would come when government policy would have a broader base than Robinson desired. The time for democracy was upon the land, but till he died Robinson would inveigh against "coarse vulgar democracy", one of whose proponents he had just vanquished.[36]

[32]*Ibid.*, May 6, 1819.
[33]*Ibid.*, Jan. 18, 1819.
[34]*Ibid.*, July 11, 1819.
[35]W. Smith, "The Reception of the Durham Report in Canada", C. H. A., 1928, p. 46.
[36]Robinson to Strachan, April 8, 1851, Robinson Papers, PAO.

Was it the back-slapping Honourable William Dickson, legislative councillor, who launched the final drive against his cousin by marriage? The day after the trial, the great patriot stood in the room where he had offered 500 acres of land to the man he had encouraged to stand up against York. He was dictating a letter to his son William, his amanuensis.[37] It was addressed to Major Hillier, secretary to His Excellency, Sir Peregrine Maitland. The trial had accomplished its purpose, he wrote with obvious satisfaction. Mr. Robinson had given his customary luminous exposition with the suavity for which he was already famous. The "Modern Reformer" had departed; Robert Gourlay would trouble the country no more. The attempt to prosecute Mr. Swayze for defamation of character was abortive. Mr. Ferguson was in jail till next sitting of the King's Bench when he would appear in York for judgement. The country could resume its accustomed tranquillity, concluded Mr. Dickson.

If Sir Peregrine had not expressed a desire to be informed of the outcome of the trial, he was certainly being served with a quick report of it. Had he asked Dickson to communicate the verdict to him, or was Dickson's zeal of his own devising? Had Sir Peregrine cried out like Henry II: Will no one rid me of this man? Had William Dickson assumed the role of the four knights who murdered Becket so that Henry II could sleep in peace? Nothing so dramatic. There is no record that the sleep of either Maitland or Dickson was disturbed or that either gave the banished Briton a second thought.

Dickson had acted in the interests of his country but Gourlay had his own opinion of Dickson's vaunted patriotism: "When a prisoner in that country (the United States) during the war, it was reported that he consoled himself, when Canada was likely to be conquered, by saying publicly that such an event would be of advantage to him, in doubling the value of his property; and his speech to the officers of the 70th regiment is undeniable."[38] William Dickson never sued Gourlay for libel. He did not dare.

Was it Isaac Swayze who was to blame? Isaac Swayze had led a troubled life from the beginning. He was always grasping at more than his due. Even many of his war claims came back with "Not Allowed" marked on them in red, and it would be his son, Francis Gore Swayze, who would repay the licence money in four annual instalments after his father's death.[39] He had a great desire for the importance that attends participation in government affairs, and he darted about with useful ubiquity. He was elected to the assembly, lost his seat, was re-elected, and lost his seat again after Gourlay's banishment. At a public meeting Swayze compared Gourlay to

[37]Dickson to Hillier, Niagara, August 21, 1819, Upper Canada Sundries, PAC.
[38]*Statistical Account II*, p. 493.
[39]August 11, 1834, Upper Canada Sundries, PAC, E383 10-12.

the instigators of the American Revolution and "that pretended patriot, Joseph Willcocks". In addition, Gourlay attributed Swayze's willingness to prosecute him to the fact that he had blocked Swayze's attempt to get some land. In March 1818, Gourlay had been visiting Thomas Dickson who had just received a letter from Isaac asking for land as a reward for his war services. Because Swayze had resigned from his command of the Provincial Artillery Drivers at a crucial moment during the War of 1812, Thomas Dickson saw no great need to reward him.[40] This caused Gourlay to write in one of his letters to the *Spectator* that it would be more to the point if the members of the assembly would forget their personal applications for land till they had performed the duties for which they had been elected. It seems unlikely that Gourlay's statement prevented Swayze from getting land. Isaac Swayze cannot be regarded as the man responsible for Gourlay's prosecution. He merely acted in character. His motive has always been termed impeccable: the service of his king and country.

There is one man left to consider, a man whose name appears neither on the indictment nor in any record of Gourlay's trial. Was it the Reverend Dr. John Strachan who was guilty of this perversion of justice? It was he who had disapproved of Gourlay first and said that his opinion of Gourlay's proposed statistical account "by some strange breach of confidence had been communicated to him."[41] His motives were pure, it would seem, because he was advancing the cause of religion, but he was not too particular about his methods. As soon as Miles had sold the *Kingston Gazette* to Macauley, Strachan wrote to his friend, the postmaster, December 8, 1818: "I am glad to find that the *Kingston Gazette* has fallen into such good hands. Although little danger can now be apprehended from the machination of Mr. Gourlay much greater danger is done to the country by preventing people from coming to settle in it. Tho we think nothing of the idle people who signed this man's petitions, this is not known at a distance and they think in England from seeing no defence of the measures of government, that the province is in a state of commotion.

"Besides a great deal of sullen discontent has been instilled into the minds of the people.

"To correct these evils I propose writing a few letters for your paper. Should I say anything that is considered libellous I will guarantee your costs.

"Keep my secret profoundly."[42]

So much for the free press that Maitland maintained had been re-established.

---

[40]*The Banished Briton*, #20, p. 224.
[41]Strachan to Harvey, June 22, 1818, G. Spragge, ed., *The John Strachan Letter Book*, p. 164.
[42]Strachan to Macauley, Dec. 8, 1818, Macauley Papers, PAO.

A few weeks later Strachan wrote again to postmaster Macauley: "It appears by the last news from Niagara that the zeal of the Magistrates of that place, which has been so long dormant has at length broken out and that they are putting the 44th of the King in force against Mr. G. This is to be regretted; for I am sure that the Atty General would have found no great difficulty in convicting him for a libel in some of his recent publications. I believe that the man is unhappy and has always from his youth been restless and turbulent."[43] Strachan often claimed that people whose political opinions he disliked were worthy of pity. When Robert Nichol returned from England with a pension that Strachan did not want him to have, it was "Poor Colonel Nichol"! Now Gourlay was "this poor man whom I have long considered with you an object of pity."[44] Strachan had not forgotten that Gourlay had published in his letter to Clark of February 18, 1818: "Do you remember my declaring, before I visited York, that I would have no introduction to this little man?"[45] Nothing can be traced to Strachan directly, and the fact that he expressed disapproval of the means by which Gourlay was brought to trial does not mean that he had not bent every effort to discredit him, for he was rarely stayed by niceties. Two years later, when the question of Barnabas Bidwell's expulsion from the assembly was being discussed on the grounds that he was an alien, John Strachan demanded: "Never mind the law; toorn him ooot; toorn him oot."[46] It was the Braxfield of Upper Canada speaking. Just as Braxfield called Thomas Muir a "demon of mischief" for pointing out to untutored people that they were being cheated of their rights, so Strachan accused Gourlay of fomenting discontent. It was the man of the right reviling the man of the left.

When no one man is to blame but a wrong has been committed, who is guilty—Maitland, Powell, Robinson, Dickson, Swayze or Strachan? "What are laws without morals?"[47] wrote Robert Gourlay who knew that the processes of law are often imperfect in contrast with "the invisible fortress of abstract truth and justice".[48] Should Gourlay have wept instead of laughed at his trial?

Gourlay had been tried and sentenced to banishment. Now it was the turn of Bartimus Ferguson, the editor who was suffering for his belief in the freedom of the press. Ferguson was tried before Gourlay but not sentenced till afterwards. The sentence passed was a £50 fine, a year and a

43Strachan to Macauley, York, Jan. 11, 1819, *ibid.* Later in the year (Sept. 26) Hillier would ask Goulburn's sanction for "a little aid to the press" from the Secret Service Fund. CO/42/362.
44Jan. 30, 1819, *ibid.*
45*The Banished Briton*, #20, p. 220.
46J. M. Dent, *The Canadian Portrait Gallery*, I, p. 107.
47*Statistical Account II*, p. 300.
48*Statistical Account I*, p. dlxxi.

half in jail, one hour daily in the public pillory for the first month of his sentence, and £500 surety for good behaviour for seven years after his release. He was not to be released till the fine and surety were paid.[49] When Ferguson promised never to offend editorially again, the pillory was omitted and the jail term shortened to seven months. The sentence was commuted by Powell acting on Maitland's order.[50]

There is no record of Robert Gourlay's trial in the minute-book of the court in Niagara (on-the-Lake) on August 20, 1819. The clerk of the court was the son of Chief Justice Powell. No person would deny the chief justice a convenience that would make a tiresome law circuit easier for him. A judge's son could learn much from accompanying his father. It is also easier to falsify records when all positions are held within a family. In England at this time there were some cities in which every important civic office was held by relatives. It was already difficult for an "outsider" in Upper Canada to procure a salaried government position. The members of the leading families claimed that they had the monopoly of brains and education; why should offices go to men of lesser attainments? Nevertheless, it was a son of one of the leading families, Henry John Boulton, who removed the record book of the Kingston trials.[51] It is not known who ordered Powell's son to omit the minutes of Gourlay's trial in Niagara.

There is a record of a Gourlay versus Swayze trial on August 25, 1819, but Gourlay had left the country by then. This is the trial William Dickson referred to in his report to Maitland when he said that the attempt to prosecute Swayze for defamation of character was abortive.[52] The Sedition Act ruled that such a suit must be commenced in prison or in banishment. If it failed, the plaintiff was to pay triple costs. Traversion of the law was to be made as difficult as possible.

Robert Gourlay had left the court in the company of his friends. For one night he slept in the house of Robert Hamilton at Queenston. The law said he must leave Upper Canada in twenty-four hours. The next afternoon he crossed to the United States on the three o'clock ferry. "I thanked God as I set my first foot on the American shore, that I trod on a land of freedom. The flow of animal spirits carried me along for more than two miles in triumphant disgust. It carried me beyond my strength, till, staggering to the side of the road, I sunk down, almost lifeless, among the bushes, and awoke from my dream to a state of sensibility and horror past all power of description. If at my trial, and so long after it, I was callous to feeling . . .

---

[49]*Kingston Chronicle*, Sept. 3, 1819; J. M. Dent, *The Story of the Upper Canadian Rebellion*, I, p. 43.

[50]Powell to Maitland, York, Nov. 17, 1819, Upper Canada Sundries, Record Book 5, A1, vol. 44, pp. 21683-21686, PAC; Maitland to Bathurst, York, Mar. 7, 1820, CO42/365, p. 98.

[51]Record Book of Assize and Nisi Prius Proceedings, PAO.

[52]*Ibid.*

214 *Robert Gourlay, Gadfly*

now, extreme relaxation resigned me to the torture of a distracted mind, feeble, doubting, irresolute!"[53]

He spent a week in Buffalo recovering sufficiently to go on. He wrote his wife that he hoped she would not hear of his banishment from the papers before she received his letter. He rested a week in Geneva and two weeks in Albany. He arrived in New York where he expected to find his brother Thomas who had left Upper Canada when his brother's political activities had made it impossible for him to get land or start a business. Thomas had left New York to go to the country with the family who was boarding him, for yellow fever was raging in the city.[54]

Robert went to see Cobbett but he received no praise for his plan for relieving England's poverty. Cobbett wrote later: "I thought he told me a tissue of lies; and besides, I did not care a straw whether his stories were true or not; I looked upon him as a sort of gambling adventurer; . . ."[55] He invited Gourlay to accompany him back to England on the same ship, but Gourlay had arranged for less expensive transportation via Montreal.

In Montreal he learned the verbal quibble upon which his trial and his banishment were based. "It was then, for the first time, that I learned that the word 'offence', used in the statute, could, at will, be applied either to an act of sedition, or to the mere refusal to obey an order."[56]

Lower Canada was at the bitter beginning of an organized opposition to the rule of the Château Clique, but Gourlay did not join it. Using Montreal as a base, he said he could have conducted "deadly machinations" against the Upper Canada administration, but machination was not his aim. It would have been easy to circumvent the law that banished him. Thousands of aliens, he pointed out later, could steal into the province, rent a tenement for six months, and attempt what they chose against the government. Though he could not see that his banishment had rendered the people happier than they were before, he would not stoop to base intrigue to reverse it.[57] He stayed five days in Montreal and then embarked for Quebec where he picked up from Mr. Neilson's bookstore an almanac and Bouchette's *Topographical Description of the Province of Lower Canada* from which he would take some statistics for the book he still intended to write.[58]

The Briton who was banished was not entirely blameless for the opposition he evoked. The intoxication of finding that Upper Canada re-

---

[53]*Statistical Account II*, p. 400.
[54]*The Banished Briton*, #22, p. 263; Gourlay Papers, PAO, Nov. 11, 1843.
[55]*Statistical Account I*, ccxv; W. Cobbett, *Rural Rides*, April 19, 1830.
[56]*Statistical Account I*, lxxvii.
[57]*Ibid.*, lxxxii.
[58]*The Banished Briton*, #37, p. 503. Bouchette has been accused of appropriating the rough draft made by William Berczy for a three volume statistical account of Upper and Lower Canada. No trace has been found of Berczy's manuscript. He died in mysterious circumstances in New York, 1813. See John André, *William Berczy*, Toronto, 1967.

sponded to his aims as Britain had not went to his head. He produced plans larger and faster than minds could take them in. Therefore he was banished, and not one street in downtown Toronto, the Upper Canadian equivalent of Westminster Abbey, was named after him.[59]

Upper Canada would never be the same again. It had been jolted into awareness of the group which in 1828 would be named the Family Compact.

---

[59]The Ontario Archeological and Historical Sites Board has erected a plaque to commemorate Robert Gourlay on the site of the jail just outside Niagara-on-the-Lake. His name is included among those engraved on another plaque on the old Town Hall in Ingersoll.

21

The Statistical Account
of Upper Canada

Robert Gourlay returned to an England that was no more tranquil than when he left. On August 16, 1819, while the hot and humid Sabbath air immobilized his mind and body in Niagara jail, members of parliamentary reform clubs were marching to St. Peter's Field on the outskirts of Manchester. Manchester was one of the large new cities born of the Industrial Revolution but still unrepresented in parliament. The orderly array of 60,000 marchers was indicative of strength and careful organization. They carried banners: Annual Parliaments, Universal Suffrage, No Corn Laws, Vote by Ballot, Equal Representation or Death. Lines of neatly dressed middle class females were scorned by lower class women who straggled after their husbands in the ranks. "Go home to your husbands," they hissed at these early suffragettes.[1] A band played music. Orator Hunt mounted a cart, waved his white hat, and his stentorian voice called for quiet.

Mounted yeomen with orders under the Gagging Acts to arrest Hunt worked their way through the dense crowd. They drew their sabres. In the melée six hundred men, women and children lay dead or wounded amidst scattered shawls and bonnets. Hunt was sentenced to two and a half years in jail. When Sir Francis Burdett censured the government, he was fined £1,000 and jailed for three months. Moderates were numb with horror. The Peterloo Massacre caused one lord lieutenant, Earl Fitzwilliam, to resign his post, unable any longer to serve an unyielding government.[2]

Sidmouth and Castlereagh, Tory leaders in the Lords and Commons, passed the Six Acts, more stringent than the Gagging Acts of 1817. It was now that a new word, "radical", made itself heard in England. Reform right to the roots of society seemed the only way. Robert Gourlay heard the

---

[1]Place Papers, British Museum.
[2]Absolute accuracy is difficult. Hammond says Fitzwilliam resigned; DNB says Sidmouth removed him.

new word when he returned to England. "I am quite a radical," he wrote soon after, "but I am one of my own sort; and up to this day, am not connected with a single individual in Great Britain in any political party. I am known both in England and Scotland, because of my peculiar opinions, and these opinions are by many misunderstood. In the foregoing pages ... you may there see that my efforts to maintain these opinions have been singular—have been such as I am proud of."[3] He was still determined to right wrongs, and the first one now was his banishment. As long as one injustice remained unredressed, he considered no one safe in Upper Canada.

He had landed at Liverpool on December 2, 1819, stayed the night with the friend in whose house he had read Melish's *Travels in America* two and a half years before, and then was off to Scotland, for his wife had abandoned Deptford Farm on October 26, 1818. How constant would Penelope have been had she been besieged by creditors instead of suitors? With her head high, Jean Gourlay had sublet to a new tenant in spite of Lord Romilly's ruling that the change was not legal till the arrival of power of attorney from her husband. The crowning hardship had come when the money paid by the new tenant for stock and crops was put in bond to secure him against seizure by the duke pending final settlement with Mr. Gourlay. The lease, with twelve years to run, she had given away for nothing.

Gourlay had not been insensible to what his wife was enduring. He had written her from New York: "What griefs I have caused you! but believe this, that no heart was ever more true than your husband's—than that of the father of your children."[4] Now in her mother's house in Edinburgh, he had to confess that he had been reduced to a piece of dirt, thrown out "from a country to which I have given myself these two years, and wherein, by the turn of a straw, I might have stood high and respected".[5] He had no home now, but he was re-united with his wife who had always been his solace, who would "not mock, but pity and think of the best means to console and cure".[6]

From Jean he learned of his father's death, October 10, 1819, and the sequestration of his estate which had paid only ten and a half shillings to the pound.[7] "Rich old Gourlay" had been laid low by the same war that had speeded his ascent. Nothing remained for the son who had been the proud young laird except the £4,000 Oliver had left to his grandchildren in Robert's marriage contract. Half the creditors were disposed to allow the money to go to the grandchildren but half were not. Two lawsuits laid

---

[3]*Statistical Account I*, p. ccxxii.
[4]*The Banished Briton*, #25, p. 325.
[5]*Ibid.*
[6]*Ibid.*
[7]*Ibid.*, p. 326; *Statistical Account I*, cccv; *Memorials of the Scottish House of Gourlay*, Rev. C. Rogers (Edinburgh, 1888).

against them in Court of Session, Edinburgh, were settled in their favour, but the money was unavailable till the sequestration was complete. To add to his burdens, George Wilson was accusing him of blackmailing his brother Alexander by threatening to reveal the secret of the lottery money.[8] As there was no use bemoaning his circumstances, he set off to see his mother in Fife.

Taking the ferry at Leith, he crossed the Firth of Forth to Petitcours where the "Kingdom of Fife" stagecoach raced its rival in a suicidal dash to the dock. Through Kirkcaldy the coach clattered to New Inn, the spot he had designated for the meeting of representatives for his ideal parliament. On he went past the mile-stones and circular turnpike stations to Craigrothie. The Druid skies were on the hills and over the house where his mother had saved every paper that mentioned her son's activities in America.[9]

He was back in Fife, the shire that had rejected his efforts at parliamentary reform, though it was the birthplace of James Wilson of Carskerdo, one of the five men who signed both the Declaration of Independence and the American constitution. He was back in Fife where the children scamper among stones that mark the graves in which are buried only a hand or a heart of a martyr. He was back in Fife, where the Earl of Kellie was again writing to London: "I beg leave to inform your Lordship that I am again candidate for the Representation of the Peerage of Scotland which I hope will meet with your approbation. . . . I am not without hopes that a friend of mine will succeed against Mr. Hume in Aberdeen."[10] The "Grand Old Earl of Kellie" was seventy-five, still rosy and vigorous, still leader of Fife though he would not be lord lieutenant for a few years yet. Soon his portrait would be hanging in the Tontine in Cupar. Sir David Wilkie would paint him with the ruby robes of the House of Lords thrust carelessly aside, his greyhound vigilant at his feet.

Robert Gourlay's place as head of the opposition in Fife had not remained empty. James Stewart of Dunearn was questioning Kellie's use of rogue money.[11] Rogue money was money provided by the 11th George I to keep criminals before trial. In 1817 the sheriff clerk demanded £1,900 for rogue money instead of the usual £300. James Stewart demanded an investigation. He found that one man had been paid from it for securing signatures for a loyal address, others to preserve the peace at an election, others to spy on those who had not damned the Thistlewood conspiracy. Rogue money had been used to buy up copies of Cobbett's *Weekly Register* "to save the people's loyalty from taint". A newly founded Edinburgh

---

[8]*The Banished Briton,* #25, p. 326.
[9]*Statistical Account I,* p. ccxxi.
[10]Kellie to Liverpool, March 4, 1820, Liverpool Papers, British Museum.
[11]H. Meikle, *Scotland and the French Revolution,* p. 232. Stewart would soon kill Sir Alexander Boswell in a duel over an article in a Tory paper defaming Stewart.

paper, *The Scotsman*, was not afraid to print, "In every county abuses in expenditure of the rogue money, though not to the same extent as among the Fifeshire jobbers will, we believe, be found."[12] Now that Gourlay was back, some of the rogue money was used to watch "Mr. Gourlay's motions and . . . for corresponding with the county member thereanent."[13]

From the papers his mother saved, Gourlay saw that his activities had been misreported. On January 3, 1820, he addressed a letter "To the Editors of British Newspapers" with corrections and a plan: "To make Upper Canada attractive for immigrants, there must be an *INSTANT PARLIAMENTARY INQUIRY . . . I PLEDGE MYSELF TO SHEW THAT UPPER CANADA, INSTEAD OF COSTING THIS COUNTRY A LARGE SUM TO MAINTAIN IT, COULD YIELD ANNUALLY A HANDSOME REVENUE TO THE BRITISH GOVERNMENT . . . THAT IT MAY BE MADE A PERMANENT AND SECURE BUL-WARK TO THE BRITISH EMPIRE, INSTEAD OF BEING A LURE TO ITS INVASION AND DOWNFALL.*"[14]

He returned to Edinburgh to find a publisher for the *Statistical Account of Upper Canada*. The House of Constable informed him that it had in its hands the manuscript of a similar book written by James Strachan. They were rejecting it, but because it had come recommended by the lieutenant governor, they could not with propriety accept one from Mr. Gourlay.[15]

He decided to walk away his disappointment. Barely two months after being reunited with his family, this compulsive "Ancient Mariner" set off on foot through Lanark and up the Caledonian Canal to Inverness. The canal he inspected with a view to making recommendations for his proposed St. Lawrence canal. What he saw of the Highlanders living in itchy turf hovels strengthened his determination to find a publisher for the book that would aid emigrants.[16]

When he went to Edinburgh in May, he found that the "Teapot Terror" of York had beaten him again. On sale in the bookstores was *A Visit to the Province of Upper Canada in 1819* by James Strachan, actually written by John. The Tory *Quarterly Review* reviewed it favourably, but the Whig *Scotsman* dismissed it as a handful of ideas under a mountain of prejudice. Gourlay counted the errors in the pages referring to him, but Strachan's work was in print and his was not.[17] Again he must go to London.

After he settled into his accustomed lodgings in Bouverie Street, he

[12]*The Scotsman*, March 16, 1822.
[13]*Ibid.*
[14]*Statistical Account I*, p. xvii.
[15]*Ibid.*, p. cxcviii.
[16]*Ibid.*, p. clvi.
[17]*Ibid.*, pp. cci, ccxiii.

applied for admission as a student at law.[18] It was refused, in spite of his contention that, though he was under sentence of banishment, he had never been convicted of a crime. He borrowed £200 from a friend.

He called at Benbow's reading room to see how many papers had printed his letter of January 3. He noticed that he was "dogged by some busy-body or spy", and "the person to whom the information was conveyed, worked himself into a passion with the thought of my association with Benbow . . ."[19] It is not known who was the McCarthy in the case, but the spy would be paid out of the rogue money.

Just as important as finding a publisher for his book was Gourlay's need to have his banishment rescinded. Only the king could revoke it. Gourlay prepared a petition to present to His Majesty, but George IV was busy. Shortly after he became king on January 19, 1820, a bill to dissolve the royal marriage on the grounds of adultery was introduced to the House of Lords. Every paper was filled with the terrible questioning: Where was the queen's bed? How bare were the queen's arms? Where was Bartolomeo Bergami? The reformers supported the queen. Gourlay pitied her for any frailties she might possess.[20]

He mailed out circulars calling attention to his wrongful banishment from Upper Canada. He asked Sir James MacIntosh to present his petition to the House of Commons, July 12, 1820. When nothing came of it, on August 18 he asked Lord Holland to present it to the House of Lords. Lord Holland said the vellum was stitched, not pasted, and he had never before seen a petition that was accompanied by printed matter, in this case a copy of the Seditious Alien Act. His lordship was busy with the queen's business, and asked Mr. Gourlay to return later.[21]

This frustration was balanced by finding a publisher in the firm of Longman, Hurst, Rees, Orme and Brown. His plan for his book was enlarging daily. In addition to the township reports and Bidwell's *Sketches of Upper Canada*, he intended to include his study of the poor in Wiltshire, his theory of immigration, the history of Upper Canada from the debates on the Constitutional Act that had created it, the story of the treatment of the early reformers and the circumstances attending his own illegal banishment. He began to work with his former vigour. By July 26, 1820, the first pages were set and by November 1, eighty-two pages were in print and the frontispiece was ready.

On December 2 he opened a letter from his wife in Edinburgh. She was very ill. He left for Edinburgh. He arrived just as she was being borne to

18Petition to the Legislative Assembly, May 18, 1846, Baldwin Correspondence, TPL.
19*Statistical Account I*, p. ccxx; *Appeal*, p. 32.
20"The queen and I," wrote Gourlay with characteristic humility, "arrived in London on the same day." *Statistical Account I*, pp. ccxxx, li.
21*Ibid.*, xxix, lv; *Appeal*, p. 18.

her grave.[22] His wife, who had supported him through all the embattled situations he had drawn her through, was gone. His mother-in-law had raised two children of her own, and taken charge of the education of five of Robert Hamilton's. Now, redoubtable woman that she was, she assumed the care of her five Gourlay grandchildren. Human charity has its limits. In her will, Mrs. Jean Hamilton Henderson left her estate to her grandchildren, provided their father had been dead for twelve months. It was the only curse the good Christian woman allowed herself.

Robert tried to resume the writing of his book where he had left it at page 197, but he could not think coherently. In March 1821 he left by boat for London but the turbulent voyage upset him. Early in April he started to walk to Wiltshire but on the second day took a coach. He visited friends, and two former labourers who told him they had opened a Sunday School for the children.[23] He walked back to London where he read that Mr. Scarlett was giving notice of intention to bring in a bill to reform the poor laws. In both the *Morning Chronicle* and parliament Scarlett quoted from Gourlay's pamphlet, *The Tyranny of the Poor Laws Exemplified*.[24]

This recognition, combined with news that a commission was coming from Upper Canada to ask for an inquiry into its affairs, had an electric effect on him. He asked MacIntosh to present a petition on immigration. He wrote to Bathurst on the subject but Bathurst said it was no longer the policy of His Majesty's government to encourage people to go to North America. Gourlay replied: "Public arrangements for emigrants were miserable, when I was abroad, . . . Nothing like a plan was laid down . . . no clear idea was formed regarding the art of settlement in the wilderness." He flared back at Bathurst who "will neither do nor let do. Like the dog in the manger, his sole object is to prevent enjoyment."[25] He added this correspondence to his book, using its pages as he had used newspaper columns before.

By December, 288 pages of the *General Introduction* to the township reports were in print. Exhausted by work and London fog, his despondency returned. The £4,000 legacy for his children was being disputed before the House of Lords. He decided to journey to Land's End and commit suicide. He wrote a friend asking him to give his watch to his son and a ring to each of his daughters. The friend advised him to throw himself into the Thames and not trouble to go to Land's End.[26]

He shook off the impulse to suicide, and finished the introduction of his book as far as page cccxlii. Despite the fact that he claimed that at times

---

22*Statistical Account I*, pp. lxxxiii, ccvi.
23*Ibid.*, ccclxi.
24*Ibid.*, p. cclxxxvii.
25*Ibid.*, p. cclxxxvii.
26*Appeal*, p. 72.

he was "fatuus", he was handing copy to the printers at a rate that would have put a lesser man under the table. Because he described his condition as "fatuus" (Latin, foolish), some historians, notably Dr. Aileen Dunham, have claimed that he was again insane as they said he had been during his trial. There was no mental aberration at this time, only an immense frustration brought on by adversity, failure and sorrow. He continued to compose one step ahead of the printers in a manner that precluded revision. On April 13, 1822, his book was published in two large, impressive, leather-bound volumes costing an aristocratic two guineas.

He had accomplished the main purpose of his book. The statistics drawn from the township reports, with the exception of one error that has been traced to Samuel Ridout in York, were correct. The description he gave of the Niagara District is the basis of all contemporary accounts, along with Michael Smith's *Geographical View of Upper Canada*, 1813. One or two discrepancies in the figures for the London District may be attributed to the doubling up of families after the Americans had burned their way through. It may be said that it was no great task to compile an account of little Upper Canada, but Gourlay proceeded in a business-like fashion. When Thomas Paine wished to estimate the number of old people in the country for *The Rights of Man,* he stood in the street counting the people who passed him in an hour, and then announced that one out of sixteen was over sixty. John Galt composed his little article, "Statistical Account of Upper Canada", in 1807, merely by questioning a relative who had visited it.[27] Gourlay's account of Upper Canada would have been complete had it not been for the small-minded interference of John Strachan in the Home District and Philip Vankoughnett and Jonas Jones in the Johnstown District.

After the reports were in his hands, Gourlay edited them for the purpose of brevity. This fact was confirmed in 1960 when a copy of an original report was found by the grand-daughter of William Chisholm, one of the signators of the Nelson Township report, who compared it with the printed one.[28]

In his *Introduction* (probably the longest in the history of publishing), Gourlay progressed from the mud-slinging of his Niagara days to a more sophisticated exposition of the ridiculous. So Mr. Pitt wished to establish an hereditary nobility in Canada! "By this day we should have witnessed a pleasant farce. We should have seen, perhaps, the Duke of Ontario leading in a cart of hay, my Lord Erie pitching, and Sir Peter Superior making a rick; or perhaps his Grace might now have been figuring as a petty parson, starving on 5,000 acres of clergy reserves."[29] So Canada was to be governed

[27]*Philosophical Magazine,* 1807; *The Autobiography of John Galt,* (London, 1833) p. 97.
[28]Hazel Mathews, "Trafalgar Township in 1817," OHS, June 1960.
[29]*Statistical Account II,* p. 296.

by an image of the British constitution! "The fancy of giving to Canada the British constitution was a good one; about as rational as to think of cultivating sugar cane in Siberia or to entertain hope from grafting a fruit twig on an icicle."[30] The anonymous gentleman who had written a pamphlet to refute John Mills Jackson's criticism of the administration was denominated Smellfungus, after one of the less attractive characters described by Lawrence Sterne.[31] Richard Cartwright never forgave Gourlay for being so well acquainted with English literature, and his son remained his implacable enemy for years. John Strachan was promoted from being a "tawdry schoolmaster" to "a monstrous little fool of a parson", but he could not bring a libel suit against Gourlay because of his own virulence in *A Visit to the Province of Upper Canada in 1819*.[32]

Gourlay's virulence erupted only after the refusal of his application for land. He had left an England that degraded the labourer and came to Upper Canada where the government countenanced the lowering of character through the struggle with unnecessary hardships in the wilderness. The orderly, systematic colonization he proposed was designed to correct both situations. In the Niagara and Western Districts he felt a strong and irritable desire for progress, a progress denied by York acting on instructions from London. Gourlay believed that a government should anticipate the needs of a people.

With that principle in mind, he assessed in his book the ideas of the fashionable political economists, not as an economist himself, but as any well-read gentleman had a right to do. He approved Adam Smith's catchword of laissez faire in trade, but not in social spheres which necessitated government planning. He put Malthus in his place with his summation: "He dwells too much on gloomy results: . . . he seems almost to doat on the idea that the condition of man is hopeless: . . ."[33] He opposed David Ricardo's manner of estimating a country's wealth by regarding the top of the pyramid—figures for rental income, balance of trade, stock exchange activity—rather than the individuals caught in the pincers of growth and change. When radical Robert Gourlay looked over a county, he saw more than fat farmers. He saw the labourers' ragged wives and children who looked to their ill-paid husbands for bread. Practical businessmen and orthodox thinkers considered the radicals as part sentimental fools and part destroyers of the established order.

Gourlay knew as well as Ricardo that rent was a substantial property, but he saw no reason why rents could not operate on a sliding scale according to the price of corn. The postwar situation was still critical. He

---

[30]*Ibid.*, p. 294.

[31]*Ibid.*, p. 336.

[32]*Ibid.*, p. xv.

[33]*Ibid.*, p. ci. Classical economics, writes Marshall McLuhan in *Understanding Media*, (McGraw-Hill, 1964) p. 11, do not take growth into proper consideration.

would adjust the income tax so that, instead of stopping it on taxable income over £150, its percentage would be increased and no one would suffer but a few who would have to dispense with "lap-dogs, footmen, and fiddlers". He would adopt the paper currency fought so fiercely by Cobbett. He would educate the poor to a position of dignity. "No half measure, no piddling plan can avail," he wrote.[34]

Gourlay's Scottish friends regarded his book as "a work of uncommon merit". Joseph Hume used it to defend Gourlay in the next situation in which he would be embroiled. (Hume was the man to whom Upper Canadian reformers wrote to explain their needs.) "He had written a very good book upon the state of Canada," he told parliament.[35]

English coffee houses bought Gourlay's book for their libraries.[36] Edward Gibbon Wakefield, soon to be imprisoned for abducting an heiress, read it in prison. The ideas he derived from it directed his endeavours as a radical imperialist, but Gourlay would receive little credit. Mr. and Mrs. J. L. Hammond in their monumental studies, *The Village Labourer*, 1760-1832, and *The Town Labourer*, 1760-1832, claimed that William Cobbett was the only man of his generation who regarded politics from the standpoint of saving the labourer from ruin.[37] They gave no credit to Robert Gourlay who since 1808 had been trying to wake parliament to the needs of the English labourer. William Cobbett did not fail to mention Gourlay as he passed through the valley of the Wily on one of his Rural Rides. He was no more inclined than Gourlay himself to praise another. He never acknowledged reading Gourlay's book.

In Upper Canada Gourlay's book was not ignored. For the first time someone had looked at Upper Canada as a whole. For the first time there was in print a summary of Gore's treatment of Weekes, Wyatt, Thorpe and Willcocks, and Maitland's treatment of Gourlay. Gourlay's supporters, called Gourlayites, formed the first political party in Upper Canada—Gourlayism—a name in use till after the 1837 rebellion.

In 1824 John Galt was asked to head the Canada Company that the government was forced to incorporate to regulate its haphazard immigration policy. Crown reserves were sold to the company that would sell the land in order to settle the still unpaid war claims. Galt used excerpts from Gourlay's *Statistical Account of Upper Canada* as promotion material for the land company that would implement a policy Gourlay was banned for advocating. Galt did not acknowledge the source of his material.

A red-haired Scotch radical named William Lyon Mackenzie founded

---

34*Statistical Account I*, p. cxlix.
35House of Commons Debates, June 11, 1824, Hansard, p. 1206.
36TPL copy is marked "John's Coffee House."
37Hammond, *The Village Labourer*, 1760-1832, p. 236.

the Colonial Advocate in 1824 to take the place of the press stifled by Maitland. Mackenzie referred to Gourlay's banishment as long after as 1835 when he composed his famous Seventh Report on Grievances. It would be Lord Durham's task in 1838 to head the inquiry Gourlay asked for in his book in 1822.

The book was read by the policy-makers of Upper Canada, whether they agreed with it or not. Attorney General John Beverley Robinson read it in London. He had been sent there to push an illiberal Union Bill (of Upper and Lower Canada) through the British parliament in order to solve the continuing difficulty of dividing revenue from tariffs. Maitland was not paying Robinson's expenses with money voted by the legislative assembly. He borrowed £3,000 from the military chest because the government coffers in the province that had no grievances were still empty.[38] Robinson was prepared to defend the Upper Canadian government against possible charges in Gourlay's book. Though some members of that government would shudder at its contents, Robinson wrote back to Major Hillier, Maitland's secretary: "It is the most absurd thing you ever saw—full of gross abuse—it contains endless falsehoods and ridiculous mistakes . . . of the township reports he had chosen to publish forming his volume of statistics, there is not one which contains a word of accusation agt. . . . the government—nothing political—not a syllable about corruption, etc."[39] In Robinson's eyes Robinson could do no wrong—therefore he found nothing accusatory in it.

Inscriptions in copies still extant indicate that James Stuart, chief justice of Lower Canada and Captain John Harris of Eldon House, London, Ontario, owned and presumably read the *Statistical Account of Upper Canada*.

The completed book was diffuse, but its organization had not faltered till the writing was interrupted by Jean Gourlay's death. During the ensuing period of adjustment Gourlay altered his original purpose of writing a brief explanation of the principles that governed his actions before he began to collect the township reports. His growing exasperation at the failure to achieve redress for the poor of England and immigrants to Canada caused his work to become an exposé as well. He attempted much, and succeeded sufficiently.

To know how much he accomplished, one has only to compare his book with similar products by J. W. Bannister, Dr. Thomas Rolph, Dr. William Dunlop, Dr. John Howison (The Traveller), Charles Stuart,

---

38*Statistical Account I*, p. cccclxix. Indeed the government was forced to borrow from the land speculators. Clark and Street to Hillier, Niagara Falls, Sept. 3, 1822, Upper Canada Sundries, PAC.
39Robinson to Hillier, London, June 10, April 22, 1822, *ibid.*

James Strachan, and above all with Mr. Charles Fothergill's *Sketch of Upper Canada*, 1822. Government-approved Mr. Fothergill, who had opposed Gourlay in Hope Township, filled half his book with heterogeneous items such as a list of the kings of England, a table of weights and measures, nine Golden Rules to render young tradesmen respectable, and the specific gravity of thorn and crabapple trees. Place these beside a book that essayed the reform of the social, legal and political structure of a country, and see how far Robert Gourlay's book, in spite of its faulty organization, will remain a monument to his memory as "enduring as a tablet of marble".⁴⁰

Gourlay had pushed his book to completion and publication. It is not known if he made any money from its sale, and he was still without employment or income. What would he do now, the man who was no longer among the favoured of Fife? He might scour books on every country in the world for more statistics to prove an erroneous theory, like Malthus. He might join Bentham's circle, for he revered Jeremy Bentham, but Bentham's circle was small, select and scornful of Gourlay's type of enthusiasm.⁴¹ He might shadow Major Cartwright, but the old warrior had already been passed over for younger men, as he nourished himself at meetings on raisins and weak gin. He might join Robert Owen, but Owen was as jealous of his projects as Gourlay was of his. So was Cobbett. When Gourlay finished working on the *Statistical Account of Upper Canada*, he sent Cobbett copies of his poor law pamphlets to show Cobbett how desperately he too had worked for the poor. In the *Political Register* of November 2, 1822, Cobbett described his reaction to the pamphlets. He bundled them up and sent them back to "this single-minded friend of mankind," advising him to try "country air, constant exercise, low diet, cooling medicines, copious bleeding, and if these will not do, a straight waistcoat or a horsewhip or both."

Gourlay was not an easy man to deflect. His constant plea in his book, "Rescind my banishment", was voiced as much for political liberty as for himself. "By calling myself a democrat, it must not be inferred that I do not love kings or peers . . . I shall never quarrel with the king, 'who can do no wrong'. I shall never quarrel with peers, while they do their duty."⁴² Part of their duty now was to heed his call for an inquiry into the affairs of one of the colonies that they administered. The gadfly was still alive.

---

⁴⁰Lesslie Papers, OA. James Lesslie came to Canada on the same boat as W. L. Mackenzie.
⁴¹Bertrand Russell, who was more of an activist than most philosophers, describes the type thus: "The Webbs and the Benthamites shared a certain dryness and a certain coldness and a belief that the waste-paper basket is the place for the emotions." *The Autobiography of Bertrand Russell*, I, p. 79.
⁴²*Statistical Account II*, p. 479.

22

Whipping a Whig

The year following Robert Gourlay's banishment from Upper Canada was an election year that the province would long remember. He had left behind him no fair weather friends. Deprived of their commissions as leaders of the militia, justices of the peace or sheriffs, denied the land they had been promised for war services, they still refused to recant.[1] Marshall Spring Bidwell was in fighting mood as he wrote to John Hunt in Detroit, May 22, 1820: "I however, will tell you what is not *dull*, i.e., electioneering. This is the Saturnalia of the mob. The spirit of Gourlayism is about, walking to and fro, and behold! many are they that are possessed. Little York trembles to its inmost puddle and lo! the chief counsellors, they that took counsel together against the man Gourlay, are compassed about with terrors!

"There is every indication of a very tempestuous election through this part at least of the Province—and they are never very eminent for quiet and decorum. Gourlayism will be triumphant at the Polls."[2]

When the election results were announced, eight Gourlayites were returned out of thirty-two. The day had gone when the reformers could be silenced by the recall of a Thorpe, the duelling death of a Weekes, or the banishment of a Gourlay. The Niagara District threw out all four former representatives—Isaac Swayze, David Secord, Ralfe Clenche and Robert Nelles—in favour of Robert Hamilton, William J. Kerr (son-in-law of Captain Joseph Brant), John Clark and Robert Randall. The other four Gourlayites elected were William Chisholm for Halton, George Hamilton for Wentworth, Samuel Casey for Lennox and Addington, and Thomas Horner for Oxford. (Robert Nichol for Norfolk and John Willson for Wentworth were often critical of government policy but were not necessarily Gourlayites.) The Cornwall country returned the book-burning and

---

[1]Beasley lost command of a regiment. Sheriffs Kerr and Merritt were dismissed. Maitland to Bathurst, York, March 4, 1820, CO42/365.
[2]Burton Historical Collection, Detroit.

horse-whipping people. York and Simcoe riding was divided between Peter Robinson, brother of John Beverley Robinson, and moderate reformer, Dr. W. W. Baldwin. The town of York elected Attorney General J. B. Robinson who would never cease to proclaim: "I prize everything anti-radical beyond measure." Formless shifting on political issues was beginning to take more definite shape.

One of the first acts of the new assembly was to repeal the Gagging Act that forbade township meetings. When it was passed, only Willett Casey dared to vote against it. When it was repealed, only Robinson dared vote for it. The Seditious Alien Act would not go down so easily. When the assembly voted to rescind it, the legislative council consistently refused to do so. A new and fanatical twist was given to the alien question when Barnabas Bidwell was elected to the assembly, declared an alien and ineligible, and expelled. His son Marshall was elected in his place, and expelled on the ground that he was the son of an alien. He was re-elected over all attempts to keep him out. The Bidwell question, following closely on Gourlay's banishment, became another justified rallying call of the reformers.

The division of duties between Upper and Lower Canada on goods entering via the St. Lawrence was still unsolved. Maitland was persuaded that a union of the two provinces might end the deadlock, and sent Robinson to London to negotiate it with Bathurst. Robinson met Burdett in fur merchant Edward Ellice's house.[3] The main proposals of the Union Bill were a £500 property qualification for assembly members in a united province, five-year instead of four-year assemblies, and two members of the executive council to sit in the assembly, not exactly a liberal set-up. When the Union Bill was being debated, Gourlay attended the House daily. Only a confederation of all the provinces, he wrote Robert Hamilton, would solve the impasse properly.[4]

He had written in the *Statistical Account*: "my work is no longer to call immigrants to go out to Canada—it is to inquire into corruption."[5] In July and August 1822, he badgered Hume, Ellice, Ferguson, MacIntosh and Brougham with sixteen letters and petitions appealing for an investigation into his arbitrary banishment and the situation in Upper Canada.[6] Some supporters—William Kerr and John Brant—had come to London for justice

[3]Edward Ellice, opponent of Selkirk's settlement, spokesman for the Lower Canadian merchants, elected 1818, introduced the bill. In Upper Canada petitions poured in against it. Its passing was prevented mainly by the last minute intervention of Sir James MacIntosh who protested rushing it through the imperial parliament "without having consulted the people of Canada". Only the trade measures were passed. Whether Gourlay had influenced MacIntosh's action can only be conjectured because of his contact with MacIntosh about his petitions.

[4]Letter cited in Gourlay's obituary by "The Whistler at the Plough", *Canadian Illustrated News*, Dec. 5, 1963.

[5]*Statistical Account I*, p. cccxvi.

[6]Petition, vol. 55, Record Office, British House of Lords.

from the province "that had no grievances". Kerr and Brant were prepared to testify on Gourlay's behalf, saying that his trial had not been legal because of his state of health, "that during trial, his conduct seemed altogether at variance with former experience; his language incoherent, and his manner indicative of partial delirium. The deponents have no hesitation in saying, that the said Robert Gourlay seemed quite incapable of defending himself against any charge."[7] One of the petitions Gourlay wanted Brougham to present asked for a new trial and the other for an inquiry into the situation in Upper Canada.

It was imperative that Brougham present the first petition before Gourlay's witnesses returned to Canada. Gourlay was in the gallery one day when the Union Bill was being debated. At the end of the debate, he hoped that Brougham would at last present his petitions. Instead, Brougham prepared to leave the House. Gourlay became so agitated at this dereliction of duty on Brougham's part that he rushed out of the gallery to remonstrate with him. It happened that Robinson was in the gallery too. He saw Gourlay rush out, and suave as usual, he wrote home to Hillier, Maitland's secretary, that he had seen Mr. Gourlay inebriated in the House.[8] Finally on the night of July 18, 1822, after he had spoken on the beer question, Brougham halted, presented Gourlay's two petitions, flung them on the table after they were ordered to be printed, and walked out. Two more petitions out of a multitude were nothing to him, but during the delay Gourlay's witnesses had returned to Canada.

On August 12, Gourlay sent a petition to home secretary Sir Robert Peel, to present to George IV on the occasion of his visit to Scotland, thinking that on the first visit in a century of an English king to Scotland, George might pardon a Scot in a magnanimous gesture. Peel passed it to Bathurst who disregarded it.

Two days later Castlereagh committed suicide. Half of the illiberal combination of Lord Sidmouth in the Lords and Castlereagh in the Commons was gone. Parliamentarians mourned the tragic end of their charming colleague, but the people of Westminster were delirious with joy as his casket passed through the streets. Cobbett danced on his grave. Lord Byron composed an epitaph:

> "Posterity will ne'er survey
> A nobler grave than this.
> Here lie the bones of Castlereagh;
> Stop, traveller, and ——."[9]

---

[7]*Appeal*, pp. xix, 32.

[8]Robinson to Hillier, London, July 19, 1822, Upper Canada Sundries, PAC.

[9]Historians are more dispassionate than Byron and Shelley but quite as uncomplimentary. A. Aspinall, *Lord Brougham and the Whig party*, p. 127: "With the passing of Londonderry (Castlereagh) and the relegation of Sidmouth into that inconspicuous place to which his mediocre abilities alone fitted him passed also the stupid and unintelligent reaction engendered by the fears of the French Revolution."

Gourlay, who would never fit into any groove, expressed sympathy for Castlereagh, for he himself had recently contemplated suicide.

He was still at a low point of hope, health and fortune. Poverty is as much a state of mind as of pocket. He was temporarily impecunious but not poor as long as he could formulate a plan for action. While he waited for his affairs to improve, he determined to adopt a cure that never failed— fresh air and exercise. This time he would combine the cure with a socio- logical experiment. He would publicize the plight of the poor by living as they lived.

With his usual flair for the spectacular, Gourlay journeyed to the parish of Wily, the only place of residence he could claim, for the lease of Deptford Farm was still technically in his name. On September 10, 1822, he pre- sented himself as a pauper to the overseer of the poor.[10] He asked for work breaking flints for the roads, the labour paupers performed for their weekly dole. Too astonished to refuse, the overseer even provided Gourlay with a hammer. Gourlay set up his umbrella by the side of the road, and there he picked, cracked and tossed, picked, cracked and tossed, till the end of the week. On Monday, when the select vestry assembled in the church to dole out the rates according to the amount of work completed, a tall gaunt gentleman lined up with the rest of the paupers. The overseer paid him six shillings and sixpence.

This was two shillings more than the allowance the parish was paying at the time, but Gourlay objected for he had cracked ten shillings worth of flints. As the overseer sought for a way to deal with the situation, he spotted a gold ring on Gourlay's hand. Paupers were supposed to be destitute before asking aid. He demanded the ring, but Gourlay would not surrender it for his dead wife had given it to him. The overseer consulted the select vestry. The select vestry called a general vestry meeting. It con- sulted a lawyer who said the ring must be given up. Finally the vestry agreed to pay their former neighbour a weekly allowance of four shillings and sixpence without requiring any labour. Taking pay without working for it made Gourlay feel like a Tory sinecurist. On October 14, the select vestry agreed to let Gourlay break flints at the rate of fourpence per load to the amount of four shillings and sixpence per week.

Once again Gourlay sat under his umbrella and picked, cracked, tossed. The fresh air and simple work soothed the man who had been born with the burden of an over-active mind, but he found that four and six would not buy an ounce of meat or drop of beer. He was reduced to a diet of potatoes. His shoes and clothing suffered wear. He returned to working at a rate that would double his pay, but was not paid for it. On November 27, the overseer rode up to him and said he would pay him no more money. On December 11, Gourlay summoned the overseer for breach of contract.

---

[10]*The Banished Briton*, #30, p. 428; #37, p. 500; *Appeal*, pp. 22, 46, 72.

On the 17th a Salisbury magistrate ordered the overseer to pay Gourlay six shillings before they left the court. The overseer then demanded the return of the parish hammer or he would summon Gourlay for theft. By December 23 he cracked six shillings worth and was paid for it, the first settlement that had been made "without disagreeable words".

Gourlay's behaviour during this pauper episode has been used as additional evidence that he was insane, for no gentleman would crack flints unless he were crazy. Far from being insane, Gourlay had prescribed his own therapy during one of the lowest periods of his life. If Castlereagh had "gone a-digging," he wrote, "he might still have been prating to Parliament". He also intended to attract enough attention to cause parliament to bring his case before the king, to force Somerset to terminate the lawsuit that was still before the court, and to publicize the poverty in the southern shires. He had just finished writing in the *Statistical Account of Upper Canada* that he had never "lost sight of the cause for which, twenty years ago, I shaped the course of my life; . . . surely, then, I must be a fool indeed, if this cause is worthless, or my schemes to advance it, are good for nothing".[11] Final proof that Gourlay was not insane came from discovering where he lived during the four months of his pauper stunt. He was housed by a Mr. Houliston of Berwick Farm, near Hindon, Wiltshire, who admired his fight for the poor.[12]

At Christmas, when the weather became too cold for cracking flints, this odd pauper popped up to London to see his lawyer about his lawsuit with the duke. He then departed on a two-month tour of the south of England to assess changes in the condition of the poor since his last tour in 1801. He found them more destitute than ever. Some friendly Whigs offered to raise a subscription for his support. He refused their offer till his name was clear of the stigma of banishment. His expenses during the tour were paid by friends whom he did not name.[13]

By March, the newspaper publicity obtained from his stone-cracking brought him into parliamentary debate. Robert Wilmot, who had read the *Statistical Account of Upper Canada*, stated that he did not think Mr. Gourlay had been used unjustly in Upper Canada. Mr. Gray Bennet, a friend of Sir Peregrine Maitland's, justified Sir Peregrine's approval of the banishment.[14] Hume was bold enough to say that if he had a reformed parliament he would impeach Lord Bathurst for his inactivity as colonial secretary, but Gourlay's sentence of banishment was not removed.

The pauper stunt was more effective against the Duke of Somerset. On March 12, after Gourlay had returned from his tour of the southern shires,

11*Statistical Account I*, p. clxxii.
12Letters headed Berwick Farm, *Appeal*, p. 69; Gourlay Papers, 1836, OA.
13*Appeal*, p. 47.
14*Ibid.*, pp. 170, 173.

a paper was brought to him at Berwick Farm with a notice that the duke was taking steps to conclude the suit.[15] It was now common knowledge that the agent who had distrained on Mrs. Gourlay had fled the country after the duke had accused him of forgery. With a feeling that at last his actions about the farm would be vindicated, he set off to visit his children in Scotland. Jean was now an accomplished young lady who read Latin and Greek fluently, and all the children were a credit to their grandmother's care. He returned to London where he wrote to both the Lord Chancellor, Lord Eldon, and Attorney General Lord Gifford, asking for a speedy settlement of the lawsuit. He was again penniless. He pawned a trunk full of clothes. After he paid for a week's lodging, he had £2: 14: 4 left. He sold the furniture from Deptford Farm. He continued to petition. On June 5 and 19, 1823, Mr. Gray Bennet presented petitions from Gourlay and some Wily parishioners asking for a parish school publicly supported and a rood of land for each household so that the people could eat if they could not emigrate. On June 27 Hume presented another petition. Gourlay wrote again to Peel and to Lord Pembroke, the lord of Wily Manor, Wiltshire. Lord Pembroke replied that as he had not studied the question of the poor he would essay no action.[16]

On January 29, 1823, the case of Somerset versus Gourlay was called in the court of Chancery. The duke was appealing against the 1816 judgement of the Master of the Rolls which implied that he had "oppressed and persecuted a private individual". This trial reaffirmed the compensation Gourlay had been allowed for late entry, but favoured the duke on the question of compensation for improvements. Gourlay appealed the latter judgement and a new trial came on in the summer assizes. Again Gourlay argued his own case. Eldon adjourned the case once and was about to do so again. The thought of further delay made Gourlay frantic. He begged Eldon for a decision, saying, "I have not a shilling. I cannot live. . . . For the last seven years I have not had a half penny of my own. I have lived on borrowed money—every shilling I had I borrowed. Last autumn, my Lord, I lived on bread and water. . . . I would have worked with my hands but for the generosity of some friends." The Lord Chancellor replied that Mr. Gourlay should have employed a lawyer. He denied Gourlay the costs of the case but he did not reverse the 1816 decision.[17] The money received from the lawsuit was swallowed up in a trice to pay debts, and so was £174

---

[15]*Ibid.*, p. 65, says Somerset appealed the 1824 judgement that affirmed the 1816 decision. It is not clear who re-opened the litigation.

[16]Gourlay to Pembroke, May 24, 1823, Liverpool Papers, B.M. 38294, pp. 241 ff.; May 29; June 3: "Come, come my lord. . . . you are not one of those bloodless things your antiquarian ancestors brought from Greece & Rome." Gourlay asked Pembroke to attend next Wily "Pay Sunday" with him. The marbles still adorn the Pembroke grounds.

[17]*Devizes and Wiltshire Gazette*, Jan. 29, 1824.

he received in the autumn of 1824 from his dead wife's share of her first husband's estate.[18]

Again he turned to petitioning. In March 1824, Sir John Astley presented three from Wily, one signed by forty females who said that the government gave more thought to slaves in foreign countries than to English women who worked for sixpence a day in winter.[19] In May he appealed to the king to rescind his banishment, but the king was not "pleased to signify any commands thereupon". In June Astley, Coke and Scarlett presented three petitions for the reform of the poor laws.

Then the Alien Act was renewed for still another year by Home Secretary Peel, now called Robert the First, King of the Aliens. An Act was passed forbidding Scotsmen exiled from Scotland for political reasons to find refuge in England or Wales. Who knew when a law would be passed forbidding exiles from Upper Canada to live in England? Gourlay racked his brain for some way to draw still more attention to his imperative need. He decided that Henry Brougham was the man he must reach, Henry Brougham, the second most powerful Whig.

He asked Brougham to present two more petitions. Daily he sat in the gallery of the House of Commons waiting for him to present them, but Brougham was busy orating about a missionary in Demerara called Smith who, though a civilian, had been sentenced to death by court martial on the basis of passages in his diary expressing pity for negroes ill-treated under British rule. Gourlay knew that Smith had been wronged, but investigation now could not bring him back to life. His exasperation with Brougham for wasting his energy on a dead man instead of a live one grew hourly.

A few months before, Gourlay had decided on a foolproof method if petitioning failed. Dr. Joseph Hamilton of Upper Canada was in London. He had bound and sealed in Hamilton's presence a packet of letters to prove that his plan had been reasoned out. The time had come for action. He could listen to debates about dead missionary Smith no longer. His best friend at his lodgings was a young law student named Bannister, for whom he had written a letter of introduction to Jeremy Bentham. On June 11, 1824, Gourlay gave Bannister the packet and set off for the House of Commons.[20]

The members were gathering in the lobby, very fine with their tall hats, surtouts and high cravats. Henry Brougham was there, likely wearing the black and white plaid trousers for which he was famous. He had seen a bolt of the material, bought it, and was too Scotch to waste any till he had

---

[18]Lethem process, Gourlay Papers, micro, OA.
[19]In Salisbury recently 2,000 poor out of 9,000 population were maintained by a rate of 12s. per pound. *Statistical Account I*, p. ccclix.
[20]*The Banished Briton*, #2, p. 5; Bentham Papers, University of London.

used it all for pair after pair of trousers. Brougham was a self-made career politician, neither landowner nor aristocrat, but incapable of being ignored. He was the same age as the tall, thin man who was entering the lobby with a lady's riding crop tucked under his arm. Brougham claimed to support the cause of reform but he regarded Cobbett's writings as part of the sedition that had to be stamped out of the country. Cobbett called him "that nasty palaverer". Diarist Creevey's name for him was "Wickedshifts".

Robert Gourlay did not go directly to the observers' gallery as he usually did. He lingered in the lobby, for he had come to see Henry "Wickedshifts" Brougham. If Brougham were doing as much for him as he was doing for Smith, Gourlay felt he would be pardoned by now. What he had in mind this day would make it impossible for Brougham to ignore the gentleman who was being wronged by his persistent neglect.[21]

Brougham had proceeded as far as the last pillar in the lobby when he felt something touch him twice on the shoulder as if a small switch were hitting against the thick cloth of his coat. Whirling around, he saw that he had been given two smart taps by a man who carried a lady's whip, a little whip with an ivory handle topped with an ivory head. The man was speaking to him with words that tumbled out of his mouth so fast as to be almost unintelligible.[22]

Brougham was startled. "Who is this man? I don't know him. What does he want?" said Wickedshifts, who had known Gourlay well enough two years before to provide him with a seat under the gallery when he was about to present one of Gourlay's petitions. He could see the little whip and the agitated expression on the face of the man, and it may be that he remembered that Prime Minister Perceval had been assassinated in the same lobby twelve years before.

Now that Gourlay had forced Brougham to pay attention to him, he was shaking uncontrollably and his words were pouring out: "You have betrayed me, sir; I'll make you attend your duty."[23] The little commotion around Brougham brought Constable Cook hurrying up. He took Gourlay's arm. "Sir, do you know what you are doing?"

"Yes," Gourlay replied, turning to Brougham. "He has injured me. My name is Gourlay. You will remember my petition of about two years ago."

Gourlay took his card case from his waistcoat pocket, bowed, and presented his card to Brougham. "Let the dead bury the dead," he said with agitation, "but do your duty by me." Brougham said, "Take him into custody. Search him. The man must be mad." (Anyone who would accuse

---

21If Gourlay's description is fact and not hyperbole, more than neglect was involved: "evidence of my cruel treatment in Niagara jail;—evidence which Brougham treacherously deprived me of before the British parliament in 1822." *The Banished Briton*, #2, p. 5.

22House of Commons debate, June 11, 1824, Hansard, p. 1206; *Appeal*, pp. lxxxiv ff.

23*Ibid.*, p. iii; *Courier*, June 12, 1824; *Sunday Times*, June 27.

Brougham of neglecting his duty must be mad!) He hurried into the
House where the Speaker was informed that an assault had been made on
the Honourable Member for Winchelsea. The constable took the lady's
riding crop from Gourlay and escorted him into a committee room off
the hall. Now that the excitement of the moment had passed, he was weak
from the effort of committing an act that he had been contemplating for
four months. He sank down with his head in his hands.

Inside the House, Brougham was relating that he had been struck by a
man who was said to be occasionally deranged. Joseph Hume rose to give
his opinion of Gourlay's action. He said he had presented poor law petitions for Mr. Gourlay who wished to return to Canada to recover some
land. Ten days before, he had received a letter that satisfied him that
Gourlay was deranged. "When sane, he was a very sensible man. His work
on Canada was creditable to his talents."[24] He had suffered great wrong in
being banished from Upper Canada. By evening Brougham consented to
Gourlay's discharge if he would apologize for his action, but discharge was
not what Gourlay wanted. He wanted justice, and the repeal of his banishment. Hume tried to persuade him to apologize but Gourlay was adamant.
Hume gave up and left, quoting an old Fife saying, "He that will to Cupar
will to Cupar". That night Gourlay lay down in the detention room of the
mother of parliaments, confident that he would be freed when the House
rose.[25]

The next day there came to pass a contingency that Gourlay had not
foreseen. George Canning, who had succeeded Castlereagh as leader of the
House of Commons, said that the man who had assaulted Brougham could
go free if he declared himself a lunatic and bound himself to the safe-keep-
ing of a friend. Gourlay had expected his action to be misconstrued, but to
gain his freedom by declaring he was a lunatic would shame him and his
family beyond measure. He said he would remain where he was till the
House rose before he would plead insanity.

From Bellamy's Coffee House in the parliament buildings where he was
taken for his meals, he wrote to the Speaker and the papers. They were
saying he was insane. He said that "only a man of uncommon strength and
toughness" could have lived through what he had.[26] He cited all the petitions he had presented during the last ten years on behalf of the poor.
"Judge me by them. . . . Their subject is the welfare of the world and the
growth of humanity."[27]

The House of Commons sent two doctors, Sir George Tuthill and Dr.

[24]Mar. 12, 1824, Hansard, p. 9556.
[25]The Earl of Kellie retired that night in better humour than Gourlay. The Earl of
   Morton had resigned and eighty-one year-old Kellie was given the coveted lord lieutenancy. At the celebration in Crail that day, the earl was in fine fettle, cracking jokes and
   slapping backs. Edinburgh *Courant*, June 30, 1824.
[26]*Caledonian Mercury*, June 24, 1824.
[27]*Evening Courier*, June 21, 1824.

E. T. Munro, to examine him. They declared that he was of unsound mind. Gourlay quoted, "Great wit to madness is allied", and sent for his own doctor. His doctor declared him sane. The first two doctors presented Gourlay with a bill of £21 for having done him the honour of declaring him mad, but he was "no sae damn daft" as to pay it.[28]

On June 25 the House prepared to rise. From the window of the Prison Room of the House of Commons, Gourlay watched the throng gathering to see the ceremony. The road from Carlton House along Charing Cross to the House of Lords was freshly gravelled.

Pall Mall had been swept so that the ladies in pink and blue with white plumes in their hats could walk at ease. Life Guards and Lancers lined the streets, and Yeoman Guards in their ruffs and red stockings stood in the Great Hall. Eight beautiful cream horses with blue ribbons and blue harness pranced between the shafts of the state coach that George IV had altered with glass at the sides and front. Inside sat His Majesty clothed in a crimson velvet mantle with a large star on the left. A velvet hat looped with diamonds and plumed with white ostrich feathers adorned the royal head. He read the prorogation speech to a crowded House.[29]

The *Sunday Times* was acute enough to compare the prorogation speech with the Throne Speech at the beginning of the session. "In Ireland which has for some time past been the subject of His Majesty's particular solicitude, there are many indications of amendment." Yes, sneered the *Sunday Times*, the Insurrection Act was renewed for Ireland's betterment. "To slavery His Majesty had not been inattentive." Yes, sneered the *Sunday Times*, the death of Missionary Smith without pardon was proof that he was not inattentive. "Six or seven million pounds had been set aside in a Sinking Fund." Yes, sneered the *Sunday Times*, money was sunk instead of being used to relieve the people. Hobhouse had asked for the repeal of the window tax, but his appeal had been negatived, 155-88. The *Sunday Times* concluded: "What benefit its solemn deliberations have brought with them to the country at large, it will be difficult for future historians to discover." Nobody banished the *Sunday Times* editor.

As the royal carriage was rolling off, Bellamy walked into Gourlay's room to tell him he was free to leave. He walked out jubilantly and turned up the Strand. Just as he was passing Coutts' banking house at Castle Court, two men arrested him with a warrant from the Bow Street magistrates. The papers reported, "It had been thought necessary in certain high quarters to have him still under coercion." Gourlay was now a state prisoner, held by the personal order of Home Secretary Robert Peel.[30]

At eight-thirty that evening, he was taken before the magistrates for

---

28*Fife Herald*, July 1, 1824; *Appeal*, p. xlii.
29*Ibid.*, p. xxiv.
30*Ibid.*, p. xxxv.

questioning.[31] Magistrate Halls allowed Gourlay to question constable Cook about the assault.

'"If you had seen me committing such an act in any other place would you have thought me insane?"

"I should not. I thought you were insane from your committing such an act in such a place." ...

"No particular violence, I believe, was used?"

"No."

"If you had seen it in the street would you have thought it nothing more than a common horse-whipping?"

"Certainly."

"And very mildly administered?"

"Yes."

"Intended more to hurt the mind than the body?"

"Yes."

"Was there anything incoherent in what I said?"

"No. I thought it odd your saying, 'Let the dead bury the dead.'"

"What! You think quoting the Scriptures a proof of insanity, do you?"

"No, I do not," said poor Constable Cook.

Mr. Halls took over. "Do you, Mr. Gourlay, call that a rational mode of proceeding?"

"As rational as things ever are. I assure you that I never was cooler in my life than at that moment. I had determined on that act four or five months before and had consulted a friend on the matter. . . . It was my intention to do it in the lobby of the House. No other place would have answered the purpose."

"Who is that friend?"

"His name is Dr. Hamilton. He is now abroad. . . . It was not by his advice but with his approbation. I wrote a letter to General Ronald Ferguson and got a friend to put his initials on it requesting that friend to recollect that I was then perfectly cool and collected . . . The friend is now here."'

Dr. James Barber, a surgeon, was called. He swore that he lodged in the same house with Mr. Gourlay, that he had known him and his father for twenty years. Gourlay had shown him two sealed letters and had been perfectly sane at the time.

A second witness was then examined by Mr. Gourlay. J. W. Bannister, student-at-law in Middle Temple, also resident in the same house, said that he had known Gourlay for three years. "He was of a singularly enthusiastic turn . . . but never unfit to conduct himself . . . he had nothing revengeful in his disposition . . . he might go farther perhaps, than men of cooler temperament would do under the same circumstances."

---

[31]*Ibid.*, ix ff; London papers.

Again Gourlay was told that if he would plead insanity he would be released on bail until brought to a proper trial. Some newspapers contributed to the fiction that the man who had whipped Lord Brougham was mad:

> "Brougham and Gourlay at first were friends,
> But, when a pique began,
> Gourlay, to gain his private ends,
> Went mad and struck his man."

"Gourlay . . . as a proof of his insanity had challenged Mr. Owen to a disputation, though what they should dispute about we know not, since their object is the same, namely "the welfare of the world; poor laws, reform and emigration connected with it; the improvement of the British peasantry, and the spread of industrious men throughout the earth; the growth of humanity and the glory of God."[32] When Gourlay still refused to say that he was insane, he was taken to the jail known as the House of Correction, Coldbath Fields. There he had the company of other "insane" political prisoners.

If Gourlay had pleaded insanity and lost his case, he might have been committed to one of the madhouses of the day. At that very time the case of Mitford versus William Benbow was ready to be tried. Bookseller Benbow was being accused of libel for having published a pamphlet entitled "A description of the Crimes and Horrors of Warburton's Mad-House at Hoxton". One of the keepers of the house said that his rule was that "if a man comes here mad we'll keep him mad, but if he had his senses when he comes here, we'll make him mad." Benbow lost the case and was assessed £500 damages. At the same time Leigh Hunt, editor of *The Examiner*, was in jail, charged with libel. To make his quarters pleasant, he papered the walls with roses, the ceiling with clouds and sky, brought in books, piano and busts.[33] It was still prison. Should Gourlay have trusted the law when so many like himself were in jail?

On July 20, he was taken before the Middlesex County Sessions. The court said it could not hold him, for no one was appearing against him. The crown attorney replied that he would oppose freeing Gourlay by the Act of 39 and 40 George 3 chapter 94 section 3 stating that if an insane person were likely to commit another offence he could not be discharged without £200 bail. Gourlay said he would weather out the situation, bowed, and left the court in custody of an officer.

Again Hume advised Gourlay to comply with the demands of the court, but still Gourlay refused. One word from Henry Brougham who had been

---

[32]*Literary Chronicle*, June 26, 1824.
[33]*London Examiner*, July 19, 1824.

the one to say "this man is mad" could have freed Gourlay. "Wickedshifts" remained silent. Not one biography of the Whig who was whipped mentions the whipping, or states that Brougham failed to open his mouth to free the man who would not say he was insane.[34]

As a state prisoner, Gourlay was assigned a fairly comfortable room in Coldbath Fields. He shared it with Mr. Tunbridge, warehouse man for Richard Carlile, that great fighter for freedom of the press.[35] Carlile was in jail too. Both men fought the Gagging Acts against seditious writings, and the Six Acts passed after Peterloo. English and Scottish papers filled their columns with editorial protests against Gourlay's treatment, and published some letters he wrote to his daughters. "My dear daughters," he wrote on June 28, 1824, "My last letter to you . . . was despatched from the House of Commons:—now I am in the House of Correction. I said that I had got by honest means in to the former House, where not one of a hundred is seated but by bribery and corruption. . . . While shaving at the window [at four in the morning] I observed a small speck become light in the village of Hampstead: some minutes afterwards, another a little to the east-ward appeared; and, in five minutes more, a third was visible under Highgate Hill: . . . At five, the buildings of Gray's-Inn-Road got illumined, . . . At this hour, with sweet liberty I could brush the dew from the grass, but I cannot get out! I cannot get out! When the clock strikes six, one of our night watchmen . . . discharges his blunderbus, and immediately a bell is rung. Presently all is bustle—unlocking and unbolting —creak-crank-clang." As he walked out on the balcony to give Tunbridge room to dress, he saw below him eight treadwheels, two propelled by women, six by men, one a sailor who cursed at "grinding the wind".[36]

The Whig papers, that had been happy to report Gourlay's every action before, were silent now that he had whipped a Whig. But for two months the *Sunday Times* kept hammering for his release, not on the grounds that he had been right in assaulting Brougham but because Gourlay had exposed the hollow pretences to activity that some members hid behind. Certainly Gourlay had tapped Brougham on the shoulder, wrote the *Sunday Times*, "but because he did so, was he to be declared insane? Was the brother of a Right Hon. Secretary, who used the whip

---

[34]Brougham, like the man who had "whipped" him, was as often worthy of censure as praise. Feiling, *A History of England*, p. 775, calls him patiently insincere. Trevelyan, *Social History of England*, p. 633, describes him at the time of his entry into the Whig Cabinet: "If his wisdom and reliability as a colleague in office had been on a level with his activity and genius as a free lance in opposition, he would have been the leading statesman of the new era; but he declined, instead, into its most magnificent oddity."

[35]*Appeal*, p. xxiv. Mrs. Carlile was in prison for two years. The main crime was reissuing Paine's *Age of Reason*. Tunbridge refused to accept any money from Carlile and lived on prison fare. Carlile could not pay his fine of £1,500; he was held as a debtor to the king.

[36]*Appeal*, p. xxiv; li, lii.

much more sharply than Mr. Gourlay, insane . . . Was a noble Duke *deranged* when he whipped an illustrious Countess? Whipping is of every-day occurrence. Go to Hyde Park in the forenoon, and to the box-lobby of an evening, and you will see the "galled jades wince" under similar chastise-ment . . . Mr. Gourlay . . . has been branded as a lunatic. . . . Now Mr. Gour-lay had given proofs of skilful pleading. . . ."[37]

The nine days' wonder of whipping a Whig soon waned. Friends sent Gourlay an elegant dressing-gown, papers, books, one of them More's *Utopia*.[38] An elderly servant who had been in his employ at Deptford Farm served him without remuneration. At one time the visiting justices were disturbed at his physical condition, for he was said to be economizing on his food in order to save enough money to get his clothes out of pawn.[39] The court would not pay to release them for him though he made the request in writing.

He did not altogether waste his time. He experimented with prison diet. In 1800 Sir Francis Burdett had succeeded in improving the scandal-ous conditions at Coldbath Fields. Now, wrote Gourlay to the governor of the prison, the prisoners eat better than the poor of Wily, though not well enough for good health. He substituted some potatoes for the bread allow-ance and tried a little meal to improve his bowels, but milk was required to make the meal palatable.

As he walked in the exercise yard, he found it distasteful to converse with such jailbirds as gamblers; but after he talked with them, he concluded that the only difference between those who speculated in stocks and those who flipped dice was that the dice-flippers went to jail and the speculators didn't. Since he was accused of being mad, he wrote to the paper, the gov-ernment would excuse him for being percipient enough to see that an influential gentleman gambler had been freed without paying a farthing fine, but the common man who rented his house for the gambling was fined £500 and served eighteen months.

Once again he prepared to defend his actions in book form. He gathered together the packet he had sealed before he whipped his Whig, all the court records of his case, a sampling of newspaper accounts both pro and con, and all the petitions he had caused to be presented for the poor and for himself. In 1826 he published *An Appeal to the Common Sense, Mind and Manhood of the British Nation*, priced at five shillings. He followed events in Upper Canada, and this same year wrote to the assembly asking why Canada was still governed by blockheads when the rest of America

37*Sunday Times*, July 11, 1824.
38Inscription in the library copy at Parkhill, Arbroath: "This book, I asked from Mr. Rob. Lyon, to keep, in remembrance of his kindness visiting me in prison—bringing his family—lending books; & zealously serving me, in every way:—Septr. 14—1827 (signed) Rob. Gourlay"
39Middlesex County Records Office; *The Banished Briton*, #9, p. 85.

chose its own governors. "The dark days of Maitland" were still upon the province.

Nineteen petitions to free him were presented in the House of Commons, either his own, or on his behalf. They came from Joseph Hume, from Forgan, Ceres, Kennoway, Ferryport-on-Craig, and Scoonie, from Wily, Hindon, and Tisbury, from Ceres again, from Longford, Stapleford, Wishford, Newton, from Dunbog, and a second one from Scoonie signed by women (were they budding Pankhursts or prospective second wives?) All were tabled. In addition Gourlay had presented six petitions for the restoration of his land, for proper administration of Chancery courts, and for his release as a sane man. These joined the others filed in the library of the House of Lords.

While Gourlay was in jail (March 9, 1825), Thomas Henderson applied to the House of Lords to be declared *curator bonis* for Robert's children. Because their father was in prison, Henderson requested authority to administer the £4,000 left them by their grandfather. This was refused. In 1827 Gourlay's mother died at Craigrothie.[40] Still he would not plead insanity to gain his freedom. "That I have been able hitherto to keep my spirits buoyant passes understanding."[41]

For three years and eight months, this singular man remained in prison, refusing to go free without a proper trial. On February 5, 1828, he gave in. There is no record of his release in the Middlesex Court records. The London newspapers carried only a brief statement that the celebrated Mr. Gourlay had posted bail with Magistrate Halls. Mr. Halls wished him a speedy journey across the border into Scotland where Gourlay said he had been corresponding with a gentleman of rank concerning an emigration project.

He crossed the Firth of Forth by ferry to Fife in the company of his friend, the Reverend Dr. Chalmers. He was full of jokes. In his hand he carried his wife's little ivory-headed riding crop. Suddenly, in the midst of an agreeable conversation, he turned and gave Chalmers' rounded shoulders three sharp blows on the heavy coat that covered them. Chalmers turned around, startled. Was his old friend mad after all? "What did ye do that for, Rob?" Robert replied, "Now I have a positive veneration for my whip for it has threshed the two greatest men of our day."[42]

---

[40]*Fife Herald*, Aug. 16, 1827.

[41]*Appeal*, p. lxxxix. A recent parallel to Gourlay's refusal to plead insanity is the case of Ezra Pound. Julien Cornell, in The *Trial of Ezra Pound*, recounts that psychiatrists certified him insane because he argued that the only way to world peace was through the teachings of Confucius. He was released after thirteen years through the intervention of fellow poet Robert Frost.

[42]M. Conolly, *Biographical Dictionary of Eminent Men of Fife*.

23

The Radical Imperialists and a Rebellion

As soon as he returned to Fife, Gourlay resumed activity. On March 24, 1828, he called a meeting in Ceres at six o'clock in the morning to form an emigration society. In two months he organized seventeen, and had a grand plan for appointing subagents in each shire. He intended also to found infant schools like the ones in Lanark and a paper called *The United Labourer*.[1]

It was with mixed feelings but no rancour that he visited Pratis, the scene of his greatest happiness.[2] He gave five hundred volumes from his library to found a public library in Ceres. Just to bring his classical library up to date, he added to the gift, "for a curiosity", a few copies of Carlile's *Republican* and the *Newgate Magazine* that he had brought with him from the more advanced haunts of London. The Ceres library committee hastily withdrew them from circulation "as they unsettled the readers . . they made such a speculation through the parish that they were likely to create great divisions in the Society." Neither Fife nor Gourlay had changed since 1808.

By November the emigration societies had fallen behind with their dues. An effective emigration scheme had to be anchored at both ends, and Gourlay was still banished from the country that was to receive his emigrants. He sent reports of his organizations to William Lyon Mackenzie, editor of *The Colonial Advocate*, the paper Mackenzie founded in 1824 to take the place of the two that had modified their ways since Gourlay's and Ferguson's stiff sentences. He was dissatisfied with Mackenzie's support. As usual, he expected every sympathizer to give first importance to his projects, but Mackenzie was as confirmed an individualist as he.

---

[1] *The Banished Briton*, #23, pp. 268-276.
[2] *Ibid.*, #45, p. 45.

In 1830 the twenty-one-year lease of Deptford Farm expired, releasing to Gourlay the £1,500 that had been placed in surety for the new tenant. In the interim Thomas Henderson was bankrupt; Mr. Adamson, the brother-in-law who had taken over Craigrothie, was bankrupt; the bank that held some of Wilson's money failed; a Mr. Clark, relative of Thomas in Canada, had failed, and a William Clark had embezzled some of the money. When it was not clear who owed what to whom, the law held it. It was still holding the children's £4,000 and the interest that was supposed to be non-cumulative. Robert's brother Thomas had gone first to Demerara and then to Van Dieman's Land to seek his fortune. His son Oliver had just sailed for Canada to become a merchant under the tutelage of John Hamilton of Prescott.

Gourlay continued his interest in the science of city planning, neglected since his student days in Edinburgh, by publishing in 1829 and 1831 elegantly bound plans for the improvement of Edinburgh. It would have been well for Edinburgh if the city fathers had accepted his plan for a tunnel through to the Grassmarket and the symmetrical twin avenues he visioned curving up the hill. When the chair of agriculture became vacant in the University of Edinburgh, he sought letters of recommendation from his friends in order to apply for the position.[3] Though even Cobbett recognized that the "malignant" Gourlay had made Deptford Farm into one of the finest farms in the county, the packet containing radical Robert Gourlay's application was not opened. In 1832 he published *Record of My Private Affairs,* introducing it with an invocation more humanist than those of the *Aeneid* or *Paradise Lost*: "Oh, may I yet be useful to my kind; and compass what I have so long desired,—the emancipation of the poor from poverty, and worse than poverty,—from ignorance."[4]

Before the poor could be emancipated, the government must change. The king was still powerful, but public opinion was becoming too strong for his ministers to ignore. In 1827 George IV ordered them to form a ministry with a majority of one against the Catholics. In spite of him, Catholic emancipation became fact in 1829. The King and Prime Minister Wellington were still opposed to electoral reform. The people of London rose against the Iron Duke as they had against the Prince Regent years before when they dipped loaves in blood and threw them at Carlton House crying, "Bread, not bullets." On an anniversary of Waterloo only iron shutters on the windows of Apsley House protected Wellington from the people.

In the 1830 election, called when the irresolute Duke of Clarence became William IV, the Tories were defeated. They had held office since 1770. In the south of England, not trusting the Whigs, the agricultural

---

[3]*Ibid.,* #1, pp. 1-13; W. Cobbett, *Rural Rides,* ed. G. D. H. Cole, 1930, p. 385.
[4]*The Banished Briton,* #13, p. 120; #17, p. 180; #23, p. 288.

workers rose against their landlords in bursts of helpless rage, the last of a long series of Peasants' Revolts. Village after village threw up leaders to demand the end of tithes, to destroy threshing machines, to duck unpopular overseers and bailiffs into ponds, and to demand wages higher than five shillings a week, or two in some parts of Wiltshire. Gallonloaf Benett of Pyt House, who had ground the faces of the poor into the dirt, was attacked and Pyt House burned.[5] The Whigs, to prove that they intended to maintain law and order while they reformed, hanged nine, transported 457 to Australia, and imprisoned many more. In the reports of the trials, the Tory papers called the men from Wiltshire more insolent than others; the reform papers called them more independent. Perhaps their sense of dignity had been assisted by Robert Gourlay who had been called insane for cracking stones beside them.

Now that the Whigs were in power, Gourlay tried to obtain employment in the Colonial Office. Lord Bathurst, who had shovelled out paupers and convicts indiscriminately, was gone. He had nearly solved Upper Canada's labour shortage with convicts as he had Australia's.[6] When the Whigs took office in November, 1830, Prime Minister Grey appointed Lord Goderich as the new secretary for war and the colonies, and progressive Lord Howick, his own son, as under-secretary. One of Howick's friends was Edward Gibbon Wakefield. When he was in jail for abducting an heiress, Wakefield had listened to his fellow convicts discussing the possibility of their forced emigration to Australia, and had read books on emigration. The one that impressed him most was Gourlay's *Statistical Account of Upper Canada* with its sensible recommendations that immigrants be concentrated in planned settlements, and that land must be sold at a high enough price to attract farmers with capital.[7] From prison (1829), Wakefield published in the *Morning Chronicle* by instalments "Letter from Sydney", pretending to be an actual settler in Australia writing home about the hardships. Lord Howick inquired about the author. Soon the young colonial theorists formed a National Colonization Society. They sought support from progressive members of parliament such as Lord Durham.

Now Joseph Hume intervened to find employment for the man he believed had been wronged. On December 28, 1830, a month after the Whigs had come to power, he rose to address the Commons on colonial land policy. In Upper Canada, he said after blasting government malpractice since the days of Pitt, desirable and accessible territory was reserved for

---

[5] In *Rural Rides*, 1826, Cobbett speaks repeatedly of the dejected assemblages of skin and bone and rags that faced him in Wiltshire. Friedrich Engels, in his *Condition of the Working Class in England*, 1845, said that peasant was synonymous with pauper in the south of England.

[6] E. G. Wakefield, *England and America*, p. 293. William IV's knowledge was sketchy: "Van Dieman's Land? Never heard of it."

[7] P. Bloomfield, *Edward Gibbon Wakefield*, (Longman's, 1961) p. 194; *The Banished Briton*, #2, p. 27.

the crown and church. "The fullest information as to the ruinous effects of these Crown and Clergy reserves is to be found in Gourlay's account of Upper Canada; but I may observe, that the author, a man who would have done honour to human nature if born under a representative government, has mixed up with much valuable statistical information an account of his own preëminent misfortunes; . . .

"I am informed that Mr. Gourlay is still unfortunate. His talents and his honesty no one will question; he was cruelly persecuted, and has had no redress; and his name is popular in Canada. The Murrays and the Twisses would have appointed a footman or a dog, rather than so honest a man as Gourlay, to some colonial office in which he might be useful. What will Lord Howick do, being still young and generous?"[8]

Gourlay sent to Lord Howick the *Spectator* in which this debate was reported, along with his opinion of the Canada Company that had been formed in 1826 to dispose of the idle crown land in western Upper Canada. Lord Howick acknowledged the receipt of the communication, but preferred Mr. Price's opinion of the Canada Company to Mr. Gourlay's.[9] Robert Gouger, a member of the National Colonization Society, sent Gourlay, over Lord Howick's frank, a pamphlet and questionnaire. Gourlay sent copies of his *Chronicles of Canada* and *Record of My Private Affairs* to both Howick and Gouger. Gourlay derived no benefit from the change in the colonial office, but Upper Canada did. On November 21, 1831, Goderich announced that land in Upper Canada was to be sold, not given away to favourites or for fees.[10] The barn door was closed thirteen years after Gourlay had said the horse was being stolen.

Now it was the turn of electoral reform. In 1831 Lord Grey prepared to make into law measures he had proposed in 1792—after thirty-nine years of opposition to a new concept by the name of democracy. Even now, when "Radical Jack" Durham proposed a clause to provide secret balloting in order to eliminate the disgraceful election riots, this "radical" measure was voted down. Only an election and the threat of the creation of fifty new peers pushed the bill through Commons and Lords. There was still the king. His ministers burst in upon him at breakfast, thrust him into the House of Lords with a half-dusted crown on his head, and put a pen in his hand. The first Reform Bill became law on June 7, 1832.

News of the passing of the bill went crashing up by coach to Edinburgh in thirty-three hours, a record journey. Glasgow, said *The Scotsman*, was one living mass of joy. In Edinburgh, on August 10, a great procession, organized by the Trades Union Council, carried a banner "To the memory of Muir, Gerrald, and others who suffered for reform." Those

---

[8]Sir George Murray was secretary for war and the colonies under Wellington. Gourlay wrote him too. *Ibid.*, #4, pp. 45, 46.
[9]January, February, 1831, Colonial Office, 500/2, 500/3.
[10]Goderich to Colborne, Nov. 21, 1831, P.A.C.R., 1935 (Ottawa, 1936), pp. 278 ff.

who did not rejoice refused to put candles in their windows till they were forced, like Dr. Chalmers, by the uproar outside their doors.

The rotten boroughs were gone. In the election that followed the passing of the Reform Bill of 1832, Robert Gourlay offered himself as an independent candidate for Fife. He stated his platform in a printed *Address to the Electors of Fife*: reform of the poor laws, separation of church and state, protection for lease holders, repeal of the Corn Laws, the framing of a bill for the right use of colonial waste land for immigrants, and regard for special Fife interests. He asked for a £500 salary, as his property was out of his hands, members of parliament were not paid, and he had a family to support. This riding of the kingdom of Fife rejected his offer, partly because of the salary, and partly because all the Reform Bill had done was to increase the electorate from one-half million voters to three-quarters. It re-elected Colonel Wemyss who had represented it for twelve years and would do so for fifteen years more. William Cobbett, who had been more fortunate than Gourlay in being elected, made the sad observation in his maiden speech: "It appears to me that since I have been sitting here I have heard a great deal of unprofitable discussion."[11] A new Poor Law was passed in 1834, but three-quarters of the people still received less than one-third of the national income.

In the summer of 1833 Gourlay was visited at Leith by William Lyon Mackenzie who was tired of promises wafted across an ocean. Since the election of eight of Gourlay's supporters in 1820, the struggle between conservative and reformer in Upper Canada had intensified. Some of the issues were solved. The militia finally received their 200 acres when the ban on lands to delegates to the Convention was lifted against Robinson's wishes. War losses were compensated. A more liberal Alien Bill was passed in 1828, though Robinson's illiberal one had first to be defeated. Egerton Ryerson was successfully challenging Dr. Strachan's claim that no nation could be truly Christian without an established church, but the bitter struggle over the clergy reserves would continue till 1854.[12] The Canada Company had built a road to Goderich before it sent in settlers, but when its secretary, Scottish novelist John Galt, attempted independent actions, he was replaced by Dr. Strachan's son-in-law, John Mercer Jones, who proved to be more compliant. When the government required supporters now, they could issue just prior to elections enough land patents to swamp the opposition.[13] Despite two successive reform majorities in the assembly,

---

11*Political Register*, Feb. 2, 1833.
12For the value of the reserves, see G. Craig, *Upper Canada, The Formative Years*, p. 136. Galt claimed that Strachan negotiated for 750,000 acres more than the original clergy reserve allotment of 800,000. J. Galt, *Autobiography*, II, p. 10.
13For the story of the 1828 election, see Julia Jarvis, *Three Centuries of Robinsons: the the story of a family* (Toronto, 1953), p. 146; for the 1841 election, W. H. Graham, *The Tiger of Canada West* (Clarke Irwin, 1962), pp. 207 ff.

the legislative council could still, and did, return its bills. Dr. Strachan was archdeacon of York, member of both executive and legislative councils, and undisputed adviser to Lieutenant Governor Maitland since his friend Alexander Wood had won a suit for damages against Judge Powell in 1823.[14] In 1829 John Beverley Robinson became chief justice, Speaker of the legislative assembly and president of the executive council.

During Mackenzie's absence in Britain, the Family Compact (so named in 1828 by Marshall Bidwell writing to William Warren Baldwin) expelled Mackenzie for a third time from the assembly. He had been elected for York County in the very Home District that had been prevented from sending township reports for Gourlay's *Statistical Account*. Mackenzie had good reason to be an ardent reformer. In the famine of 1800 that changed the course of Gourlay's own life, Mackenzie's widowed mother sold her tartan to buy barley meal for her boy. "Canadians!" wrote Mackenzie in the *Colonial Advocate*, "You have seen a Gourlay unlawfully banished; a Thorpe persecuted and degraded; a Randall cruelly oppressed; a Mathews hunted down even to the gates of death; a Willis dragged from the bench of justice, slandered, pursued even across the Atlantic by envy and malice, and finally ruined in his fame and fortune and domestic happiness; you have seen a thousand other less noted victims offered upon the altar of political hatred and party revenge. . . .

"We come, at last, to the leading question, What is to be done?"[15] When John Beverley Robinson read Mackenzie's barbed articles, he raged that another reptile of the Gourlay breed had sprung up.

It was hard to loosen the hold of the Family Compact. Nevertheless, the colonial office sent an order for the dismissal of Attorney General Boulton and Solicitor General Hagerman for their part in Mackenzie's expulsion. Mackenzie asked Gourlay to return with him to Upper Canada, saying that the reformers would pay him for his leadership. Gourlay said that the people must first settle their account with him before he would begin a new page. His private opinion of Mackenzie was that he lacked stability. He showed Mackenzie and his wife around Edinburgh Castle and Holyrood and said goodbye to them.[16]

He was still looking for employment so that he could gather his family together. Jane, as eldest daughter Jean now called herself, was governess at

---

14When Judge Powell discovered that Alexander Wood had been named to replace Robinson on a Commission of Enquiry, he objected because of Wood's reputation as a pederast. Wood's winning the suit for defamation of character marked a triumph for Dr. Strachan. Powell's fine was set at less than £500 so that he could not appeal to the king in council.

15"Appeal to the Electors of York on the occasion of the proposed second expulsion of Mackenzie from Assembly", 1832. See Margaret Fairley, *The Spirit of Canadian Democracy* (Progress Books, Toronto), p. 83.

16*The Banished Briton*, #2, p. 20. Mackenzie was expelled for the third time while he was in Britain.

North Park, near Kirkcaldy. Jessie was living with Aunt Patton, relative of Sir Henry Torrens. She laughed unmercifully at the latest religious tract that serious Jane had sent her. Catherine was seventeen and living with grandmamma Henderson and Uncle Thomas, now tenant at Newton of Wemyss. Jane wrote to Catherine: "I think it would be an improvement in the arrangement of your time if you would sew an hour or two between your French and Italian studies and save Latin till after dinner. . . . Let it also be firmly . . . impressed upon our minds, that . . . if our object really be truth, [we should not] twist the words of Scripture . . . The passages which naturally occur to us in connection with those under consideration are such as these Hebs. VI 4-8. X 26-30. I John V.16. . . ".[17] Gourlay asked Jane to influence the trustees of the £4,000 to release some of the accumulating interest so that he could go to the United States to try his luck. Jane would not do this, but she offered her fifty-five year old father her half year's salary as governess as soon as it was due. On December 5, 1833, Robert Gourlay left Edinburgh, and on December 22 arrived in New York.

Almost immediately Gourlay received a letter from Mackenzie soliciting his support. Again he refused. Because he wanted access to his property at Dereham, he wrote to whomever might intercede for him: William IV, the Duke of Wellington, Sir John Campbell, British solicitor general; James Buchanan, British consul in New York; cousin Alexander Hamilton at Queenston; William Chisholm in Nelson township. Though he maintained that he would be justified in entering Upper Canada and regaining use of his property by force if necessary, force was not his way. He joked about being an invasion of one.[18] No one was more loyal to Britain than he. When Hume inveighed against "the baneful domination of the mother country; and the tyrannical conduct of a small and despicable faction in the colony", no Tory was more critical than Gourlay. Had Hume listened to him and his petitions more attentively, Canada would not now be a disgrace to the British empire because of political unrest, he wrote.[19]

He planned to write a history of Upper Canada while it could still be obtained from living witnesses. He printed a broadsheet asking for support from free Americans for a wronged Briton. Whether he received any money is not known, but this appeal made his presence known to Mr. Houliston, the man who had sheltered him during his flint-cracking in Wiltshire. Houliston immediately offered the hospitality of his new home

---

[17]Jane to Catherine, North Park, July 30, 1832, Gourlay Papers, PAO.

[18]Those who look for contradictory statements in Gourlay's writings will have no trouble in finding them. In a letter to John Neilson, Boston, June 3, 1848, he wrote: "Let me tell you that in December 1837 I felt confident that I could land at Malden with sufficient force to sweep Canada from end to end of British rule and invest Quebec before Fool's Day. . . . fear did not in the least restrain me." The 1837 utterance is nearer the truth than the braggadocio of 1845. *The Banished Briton*, #37, p. 504.

[19]*Ibid*., #2, p. 16.

near Willoughby, Cuyahoga County, Ohio. Gourlay accepted, saying that his daughter Jane would imburse him for his care in case of his sudden death.[20]

It was hard for an active man to be idle. He decided to compile a statistical account of Ohio and asked the legislature to authorize his work. In December 1835, he wrote a letter to Thomas Clark beginning "Monsters", by which he meant Upper Canada for making him offer his services as statist to a foreign country.[21] In January 1836, he offered his services on a commission of inquiry to the new lieutenant governor, Sir Francis Bond Head.[22] Inquiry was not Sir Francis' prime concern. He disliked "that low-bred antagonist, democracy" as much as Robinson and Strachan.

To fill in time, Gourlay wrote in his journal. He wrote to his family. He longed to know if Oliver had landed safely in Singapore, after trying an inhospitable Prescott. He wrote to Jane, the pious daughter: "Oh, that it were with me as in months past, as in the days when God preserved me, . . . as it was in the days of my youth . . . when my children were about me. But now they that are younger than I hold me in derision whose fathers I would have disdained to set with the dogs of my flock."[23] Jane had no sympathy with suffering Job. She answered: "Oh my dear father, that you would see the utter vanity of attempts to work good in the world."[24] He wrote to Catherine: "I have killed care with gardening these last four weeks . . . ". Catherine replied practically: "I wish you would follow my plan which is not to mind parliament and its motions, but to go and clear a nice little farm in the woods, and send for two of your daughters to keep house for you."[25] Father was not the type to bury himself in the woods. "Sir F. B. Head has gone to Loggerheads with the provincial rogues and fools," he wrote on May 28, 1836. "All is commotion there and no doubt an inquiry into the state of the province which I recommended 18 years ago must now take place."[26]

Unable to restrain himself any longer, and thinking that the growing strength of the reformers would protect him, he crossed the Detroit river at Sandwich on September 17, 1836, into the land he was forbidden to enter on pain of death. He stayed with Mr. Asken for three days, and then set out for York where he intended to appeal for pardon from the lieuten-

---

20April 30, 1836, Gourlay Papers, PAO.

21*The Banished Briton*, #23, pp. 281-4.

22*Ibid.*, pp. 2-10.

23Willoughby, Ohio, April 2, 1836, Gourlay Papers, PAO; Job 29: 2-5; 30-31.

24*Ibid.* Jane herself would soon embark on a philanthropic course. She established a home for girls in connection with St. Johns Church, Edinburgh. In 1879, at the age of seventy, she travelled to South Africa to help a recently established mission. She died the following February 1880, and was buried at Umtata, Transkei.

25Willoughby, Ohio, April 19, 1836; Nov. 23, 1834, *ibid.*

26Willoughby, Ohio, Feb., 1836, ibid; Gourlay had expected Head to be the liberal poor law commissioner. *The Banished Briton*, #2, p. 9.

ant governor. He expected to make a triumphal progress through the restless Talbot settlement, but instead, somewhere in the vicinity of Chatham or Amherstburg he "judged it prudent to pull up".[27] His Majesty's magistrates were ready to perform their duty. Gourlay returned to Detroit and boarded an American ship for Cleveland.

In Cleveland, in a memorandum to Mr. Prince, member of the Upper Canadian legislative assembly, on emigration societies, he called himself for the first time the Banished Briton.[28] Here he printed the first number of a publication in which he would relate by instalments the story of his trials. It was entitled *The Banished Briton and Neptunian being the Record of the Life, Writings, Principles, and Projects of Robert Gourlay, Esquire*, now Robert Fleming Gourlay. While he was gathering the letters and newspaper accounts that formed the verbatim record, he was stricken by erysipelas, a nervous disease accompanied by fever and intense itching. By June 1837, when eighteen-year-old Victoria became queen, he could walk across his room only with the aid of a cane.

The political situation in British North America was worsening. Lord Glenelg, the current secretary for war and the colonies, had been convinced by Mackenzie's 500-page *Seventh Report on Grievances*, 1835, and M. S. Bidwell's *Address from the Assembly*, April 15, 1835, that there was a strong reform movement. He instructed Bond Head to adopt conciliatory measures. Instead, Head dissolved the assembly in 1836 when it refused to vote supplies and campaigned personally on a platform of loyalty or disloyalty. Fifteen hundred land patents were issued just prior to the election.[29] A Toronto meeting called to support Papineau against the Château Clique displayed a banner:

> "Sweep from our shore oppression's shame, . . .
> Both Hume and Brougham have proved vain . . .
> Canadians! to the fight again."[30]

In Montreal an Address of the London Workingmen's Association was answered, reminiscent of the days of the London Corresponding Society. In Nova Scotia Joseph Howe proclaimed in an 1836 election speech: "All we ask for is for what exists at home—a system of responsibility to the people. . . ." Mackenzie, expelled five times from the assembly, elected mayor of Toronto, made a "Declaration of the Reformers of the City of Toronto to their fellow Reformers in Upper Canada": "Government, is

27*Ibid.*, #4, p. 48.
28*Ibid.*, #1, p. 14. He added his mother's maiden name, Fleming, as his mail was being confused with another of the same name.
29F. Landon, *Western Ontario and the American Frontier* (Ryerson, 1941), p. 49.
30Quotations here from M. Fairley, *The Spirit of Canadian Democracy*, pp. 43, 26, 42, 32.

founded on the authority, and is instituted for the benefit, of a people; when, therefore, any Government long and systematically ceases to answer the great needs of its foundation, the people have a natural right given by their Creator to seek after and establish such institutions as will yield the greatest quantity of happiness to the greatest number." In March, 1837, the Russell Resolutions authorized the governor to ignore the objections of the Papineau party and appropriate treasury funds that the assembly refused to vote.

Gourlay watched it all from Cleveland. Towards autumn, when he had recovered from erysipelas, he determined to seek redress by entering Upper Canada at Niagara where he had more friends than in the western section. He was too slow. In November Lower Canada revolted, and on December 5 Mackenzie armed against the immovable government at Montgomery's Tavern on Yonge Street. Because of Gourlay's longstanding grievance, he was expected to join Mackenzie. On December 26, 1837, he received a letter from Buffalo signed Mackenzie asking for his help in the cause of liberty.[31] Because his name was spelled "Goorley" and the diction was crude, he considered it a forgery perpetrated by his enemies. He wrote Mackenzie asking for an explanation and enclosing a copy of *The Banished Briton*, the last thing in the world Mackenzie would have time to read on Navy Island. On December 27 he wrote to Lieutenant Governor Head: "With sorrow and shame have I read of rebellion in the Canadas:—sorrow for the poor deluded people—shame for British rule."[32] In spite of this, the rumour spread that he was raising 10,000 volunteers to march against Canada. Van Rensselaer despatched an aide to solicit his support, but Gourlay would have nothing to do with armed rebellion.[33]

At a Patriot's meeting in Cleveland on January 2, 1838, he refused to help raise troops and reported their plans to Head, for he was a loyal Briton. This time Bond Head wrote a personal note of thanks. Again on February 10 he warned Head that 3,000 men were preparing to cross the ice to Canada from old Fort St. Clair and Sandusky.[34] Again he confounded those who had read into his actions only what they wanted to understand.

In March, when Upper Canada's new lieutenant governor, Sir George Arthur, was sentencing Lount and Mathews to death, and counselling decimation for the rest of the captured reformers, Gourlay returned to the Houlistons in Willoughby, plagued with erysipelas, penniless. A letter from Jane was being held in York for want of a dollar postage. When he received the letter, Houliston was given the twenty pounds Jane had sent. Gourlay drew on a New York bank for some of the interest on the £4,000

31*The Banished Briton*, pp. 14, 16.
32*Ibid.*, p. 12.
33*Ibid.*, #4, p. 48, quoted from *Upper Canada Liberal*, St. Thomas.
34*Ibid.*, p. 14.

bond that Adamson was holding. At that moment, Oliver was writing his dear sister Jeannie about the same money. He was now in Port Philip, Australia, where he was certain he could become a success if he only had a little capital. "I am now nearly 27 years of age & have as yet never received a farthing to set me a-going. . . ." He had sold tobacco ordered from Sydney at 75 per cent profit. "Now what I want is this that the said money should be divided now— . . . That my father should receive a sum sufficient to purchase an annuity of £160—and the balance to be shared equally amongst us five—. . . . ."[35] At the moment the money was beyond the reach of either because of lawyer Clark's embezzlement. Thomas Henderson had expended large sums to improve Newton Farm only to have it converted by its owner into kept cover, "the utter destruction of the crops that afterwards took place annually, being then unlooked for, and inconceivable."[36] Henderson was unable to pay the back taxes on the land in Dereham Township that had never been registered in Gourlay's name.[37] In 1835 the sheriff made the land over to Thomas Clark whose will in 1837 empowered his executors to sell it on behalf of his sisters, his beneficiaries in Scotland.

In order to be pardoned and gain access to his land, Gourlay determined to petition Lord Durham who arrived in Canada at the end of May 1838 as a one-man commission of inquiry into the conditions that had caused the rebellions. When Gourlay received no reply to his communication, he decided to reach Durham in person. Durham was moving fast. By the time Gourlay reached Cleveland, Durham was already at Niagara Falls, where he created great consternation among the Family Compact by drinking a toast to the American president at a dinner accorded him in Buffalo. It was a gesture of world politics beyond the ken of chauvinists. Gourlay dashed into Niagara, but Durham had left for Quebec. Replenishing his purse from friends, and with his head free from a noose, Gourlay set off for Quebec. On August 11 he was received at Government House by Colonel Couper who informed him that His Excellency had received no communication from him. Gourlay gave Couper copies of the records he had enclosed in the original packet.

He returned to his hotel, and being exhausted from his journey, called for a candle at eight o'clock so that he might retire. Just then he was told that a gentleman wished to see him. It was Edward Gibbon Wakefield who with his fellow radical imperialist, Charles Buller, had accompanied Lord Durham as adviser. It was the same man who had read Gourlay's *Statistical Account of Upper Canada* seven years before and quoted from it in his

---

35Port Philip, Australia, Oct. 21, 1838, Gourlay Papers, PAO.
36*Ibid.*, Nov. 23, 1838; April 6, 1841.
37Lot #3, fourth concession, Dereham Township, Oxford County Record Office, Woodstock, Ontario.

pamphlet on *Colonization*. Wakefield and Buller would be responsible for a good part of the famous Report that Durham would issue recommending measures to settle the unrest in the Canadas. Colonel Couper had told Wakefield that Robert Gourlay had called, and Wakefield had hurried out to thank the man who had inspired his actions. With gentlemanly deprecation Gourlay replied that his book, because of distracting circumstances during its writing, was only an imperfect presentation of his views. "Nevertheless," replied Wakefield, "a colony had been established on your principles, in Australia."[38]

This might have been the crowning moment of Robert Gourlay's life, the vindication of his seemingly ineffective career, but he did not shout for joy. He was sixty-five years old, his ancient name tainted by a sentence of banishment. He was not interested in the back-room boys who had accompanied Durham to Canada for only the representative of Her Majesty could serve his purpose. He did not rejoice that the poor of Britain were being well settled in Australia. He only mourned that he had never been able to institute his own grand plan in Canada.

Durham paid no attention to Gourlay's submission. He was ill and proud. On his return to England, he was hounded by the same implacable Brougham who had said no word to free Gourlay. As for Gourlay, he returned to the Niagara District where he was confined to bed with erysipelas with time to compose a scornful poem entitled *The Durham Ox*.[39]

He talked to the people who now repented for their part in his banishment. Charles Fothergill said that if Gourlay ever returned to the province he would go down on his knees asking for forgiveness. They settled for a handshake in a Niagara hotel in 1839. John Holmes, one of the jurors at the Niagara trial, had asked his forgiveness in 1822 in London, saying he had never meant him any harm. Seven members of the Church of England, whom he met on a steamer, insisted on giving him some money to apply to any benevolence he wished as a token of their regret for his treatment. In May 1839, he met Dr. Strachan face to face for the first time, "and the same day, he bowed himself into my forgiveness". In 1844 he would spend a day with former editor Stephen Miles during which they both forgave each other.[40]

No one hanged the Banished Briton, for he had earned the right to stay in Upper Canada because of his help during the Mackenzie rebellion.

38*The Banished Briton*, #2, p. 27; P. Bloomfield, *Edward Gibbon Wakefield*, p. 194.

39*The Banished Briton*, #2, p. 26. Cobbett, *Rural Rides*, p. 698, mentions taking his sons to see the famous Durham Ox.

40Fothergill, *The Banished Briton*, #37, p. 498; Holmes, *Statistical Account*, cccxxxiii; Dr. Strachan, *The Banished Briton*, #26, p.352. Bloomfield, *Edward Gibbon Wakefield*, quotes Duncombe's meeting with Gourlay, October 8, 1838, and raising a small subscription for him "as he seemed poor and ill-used". He still carried his wife's riding crop.

Nevertheless the stigma of an unremoved sentence rankled like an old wound. In still another *Address to the Resident Landowners of Upper Canada*, St. Catharines, January 10, 1839, he sounded the same complaint: "What did Upper Canada gain from my banishment, and what of good is now to be seen in it?"[41]

---

[41] *The Banished Briton, #2*, p. 31.

24

The Stone that the Builder Rejected

Though no one had officially strung him from the gallows, many people still regarded Gourlay as a security risk. When the rebellion broke out, a Mr. Thompson, formerly from Indiana, was arrested for having in his possession a copy of *Principles and Proceedings of the District of Niagara*.[1] Gourlay tried to locate copies of the four pamphlets he had published in Upper Canada in 1818, but none was to be found. They had all been destroyed or hidden. One day in September 1839, he was driving around the Niagara peninsula with Dr. Woolverton (a post-loyalist American liberal) to publicize a "Durham Meeting" at Thorold. On top of the escarpment near Grimsby, they passed the farm of a Scot whom Gourlay said he must meet. When Andrew Muir saw who was in the wagon, he struck at Gourlay violently. Dr. Woolverton whipped up the horse but Muir threw stones at them as they careened away.[2] Another man tore down a placard Gourlay had put up in a tavern because it contained the words *the rights of man*.[3] Gourlay instituted actions against both men as a step towards ending violence during electioneering.

He was still trying to have his banishment rescinded. The previous spring, when he had been ill with erysipelas at Major Secord's house at St. Davids, he had been visited by Richard Woodruff, member of the legislature. Woodruff promised to present a brief from Gourlay to the assembly asking that a committee be appointed to investigate the circumstances of his banishment.[4] Finally Woodruff sent him word that the committee of Thorburn and McMicking was ready to report that "waiving the legality of the judicial proceeding . . . however far he might have deviated from the law of the land . . . your Honourable House would render an act of justice to the Petitioner by addressing His Excellency [Sir George

---

1 *The Banished Briton*, #30, p. 427.
2 *Ibid.*, #3, p. 34. Thompson had fought in every battle on the Niagara frontier, War of 1812.
3 *Ibid.*, #7, p. 70.
4 *Ibid.*, #5, p. 50.

Arthur] . . . to enable the said Robert Gourlay to return and reside in the Province if he shall think fit to do so."

This was far from being the exoneration Gourlay expected, for in addition to the reversal of his banishment, he wanted his trial declared illegal and unjust. He caught a steamboat for Toronto. On April 29, 1839, Thorburn, seconded by Woodruff, moved that the report be adopted. Immediately Gourlay's enemies moved various amendments. The *Mirror* reported dramatically on May 3, 1839, that Attorney General Robinson, member for York, "appeared like a ship floundering in a heavy sea; and when all hope of shielding the character of 'native malignancy' had failed, he sunk on his chair as pale as a corpse; and, as we then thought, his last dying words, when he turned his eyes to the speaker [who had just cast the deciding vote in favour of Gourlay] were, 'I am deserted by my friends.' " The solicitor general coaxed, wheedled, implored, menaced, turned "pale as a dolphin" from despair, or "red as a turkeycock" from frustration. Gourlay's supporters compared his persecutions with the cruelty of the times of Nero, Caligula or Domitian, and recalled that every man "who had the honesty to reprobate the unconstitutional act, or express sympathy for the sufferer, was marked out as a victim by the tools of power, and denounced as a Gourlayite and a traitor".[5] After a seven-hour battle the committee's report was carried.

When Gourlay arrived in Toronto the morning after this victory, he did not congratulate his friends. He blamed them for having him *pardoned* when he had committed no offence. Though they said he was making them look foolish, he had a placard printed and posted asking permission to appear before the Bar of the House to explain his case himself. Next day he sent each member a copy of the first issue of *The Banished Briton*, the first of a long series of presentations. In the evening he went to observe the House give second reading to the report he disapproved. His impulse was to jump on the floor and take the blockheads by the throat, he said in the picturesque language of the day, but remembering that he "had got three years and eight months imprisonment in London for the mildest breach of privilege ever perpetrated", he restrained himself.[6] On May 9 William Hamilton Merritt waited on Sir George Arthur who was pleased to extend his mercy to Mr. Gourlay according to the recommendation of the House. Everyone was satisfied that justice had been done except Gourlay.

In February he asked for an interview with the governor, Mr. Poulett Thomson, later Lord Sydenham, who had been sent to implement the

5*Ibid.*, p. 56.
6*Ibid.*, p. 54; #7, p. 72. *The Monkey War*, a poem parodying *The Ancient Mariner*, was written at this time to prove, by ridiculing Mackenzie, that Gourlay was right and Mackenzie wrong. For an explanation of its references, see W. R. Riddell, "Humours of the Times of Robert Gourlay," *Transactions of the Royal Society of Canada*, XIV, p. 82.

union of the two Canadas. Thomson was busy with bigger problems. Again Gourlay was confined with erysipelas. In the spring he went to his land in Dereham. Though there was the possibility that Henderson might pay the back taxes and evict him after he made improvements, he risked development, for the Canadian government had at least decreed that land would be forfeited if settlement duties were not fulfilled. He divided the land into lots, built a log house, cleared land for a garden, erected fences, planted fruit trees, took a kitten from the Lossings to be a mouser, and in the evenings took his ease on the long verandah. He explored his land minutely. "The south half of the township contains the finest pine trees I ever saw, trees upward of 150 feet high. The north half, on which my lands are situated, has a growth of maples and beech, with some ash, elm, hickory, and basswood, averaging nearly 120 feet in height. An elm tree on my land and near my log-house, was by actual measurement in 1841, sixteen feet around, five feet above the ground; seemed to be nearly the same size upwards for forty feet; and, at a mile's distance, was seen towering over the neighbouring trees; say, in all, 130 feet in height."[7] He corresponded with Samuel Street with regard to back taxes.

In 1841 he travelled to Kingston to observe the first session of the joint legislature of the two provinces, Canada East and Canada West. He intended to lay another petition before the House and be on the spot to prevent another fiasco. On August 25, 1841, his petition was referred to a committee of Hon. John Neilson, Captain Steele, D. F. Viger, and Dr. William Dunlop, chairman. On September 11 its report was read: "He was banished from the Province for life, under pain of death should he return, his alleged crime being that he neglected to quit the country upon the order of a single Magistrate, (two legislative Councillors), acting under an unjust construction of an unconstitutional statute, most illegally exercised. ... Your committee are further of opinion, that his trial and sentence, when in a state of bodily and mental weakness, from the sufferings which he had undergone, which prevented him from defending himself, was unjust, unconstitutional, and cruel."[8] Three members of parliament, William Hamilton Merritt, Edward Thompson and David Thorburn, then gave evidence that Gourlay's intentions in calling meetings in 1818 had been patriotic and loyal and that during his trial in Niagara his speech had been incoherent, his looks wild, and his general attitude unconsciousness of what was taking place. The committee recommended that his sentence of banishment be null and void, that he be compensated for losses sustained, and be given some allowance to defray his expenses while in attendance before the legislature defending the rights of a British subject.

The discussion was less acrimonious than it had been in 1839. Gourlay

[7]*Emigration and settlement on wild land*, p. 14.
[8]*The Banished Briton*, #5, p. 57; #6, pp. 60, 67; Jls. Leg. Ass., Aug. 25, 1841.

was informed that he had but one enemy in the House, John Solomon
Cartwright, son of Richard "Smellfungus" Cartwright.[9] Sir Allan MacNab
regretted Gourlay's sufferings but claimed that the present proceeding was
unconstitutional. Captain Steele considered that the country was indebted
to him for his conduct during the Navy Island affair. John Simpson, the
"unlettered yeoman" who had opposed him in 1818, said now that it was an
unheard of thing to prosecute a Scot as an alien. Mr. Hincks considered that
they ought to spare Gourlay a pension from the £5,000 fund that had just
been voted.

The *Kingston Herald* declared that the report was "perspicuous, com-
prehensive, and trumpet-tongued. It will go down to posterity as the best
commentary on compact government. With this report before him, the
future historian of Canada will make marvel, not that an ignorant people
rebelled, but that they so long quietly submitted to outrages against law,
decency, and common sense. He will drop a tear over the fate of Lount and
Mathews, Moreau and Von Shoultz, while his blood freezes within him to
think that a human being once existed who, callous to feeling, could confirm a hard sentence, and, unmoved by the cries for mercy from thirty
thousand of his fellow-subjects, could consign to an ignominious death the
companions of his youth, who were unquestionably honest and brave.—
Aye, and that this man was not only the accuser of Gourlay, at Kingston
and Brockville; but, after two honorable acquittals, could have the audacity
to stand up to revile and condemn him, feeble and distracted in the Jail
and Court House at Niagara! ! ! "[10]

Again, on September 16, Dr. Dunlop summoned all his eloquence to
answer objections to the adoption of his report. They said that Gourlay had
been banished legally! Every one knew that Mr. Gourlay had been an
inhabitant of the province for more than two years. How was this got over?
Why, by a most lawyer-like and quibbling construction as to the meaning
of habitancy. Dr. Dunlop said that he had been accused of using strong
language in the report. He said no language was too strong to denounce
tyranny and oppression. What could compensate a man for twenty-two
years of complicated misery, for imprisonment, for legal infamy, for ruined
health, and for a shattered constitution? It was argued that the petition
was phrased improperly and should not be accepted. The Tiger urged:
"Considering the sufferings of the petitioner, it would be a hardship for
the house to deal strictly with the petition. It was only last year that we
took from above his head the sentence to hang him. The Governor had
the power to order the sheriff at any moment to hang him up. . . ."[11]
Dunlop's report was signed the next day by Lord Sydenham who was to

9*The Banished Briton*, #6, p. 67.
10*The Banished Briton*, #6, p. 62.
11*Ibid.*; Jls Leg. Ass., Sept. 28, 1841.

decide what compensation Gourlay should receive. Three days later Sydenham died as a result of a fall from his horse two weeks previous. It was October 26 when the interim administrator, Sir Richard Jackson, recommended that Gourlay be awarded a pension of £50.[12]

Gourlay was incensed. A common soldier would have been given that amount. He was a scion of one of the oldest families in Fife, and till he was ruined by an illegal trial had enjoyed not less than £500 per annum. He gathered the record of the latest proceedings in his case to present to the new governor, Sir Charles Bagot. Bagot's sailing was delayed. Gourlay travelled restlessly to Avon Springs for warm salt water baths to cure his bad leg, then to Kingston, to New York, back to Dereham in the summer of 1842, and to Kingston where the Dunlop committee re-opened his case before Sir Charles Bagot. On October 10, 1842, Bagot endorsed the award of a £50 pension. Gourlay had accepted one payment of £50, pending adjustment. It was the last money he took of the pension that he considered belittled him.

He busied himself circulating a poster from which he hoped to obtain a list of subscribers for a paper to be called *The Commonweal and Canadian Farmers Joint Stock Press*.[13] He was still trying to organize farmers, but Canada had not even repealed the Combinations Act.[14] He reprinted *Chronicles of Canada* in St. Catharines, 1842, as the beginning of a history of the province, he being the central figure, of course. He travelled to Providence and then to Boston where he resided in the Marlboro Hotel which was frequented by members of the Massachusetts legislature. He published a plan for the improvement of the city and donated an enlarged daguerrotype of himself to the Athenaeum. Time was heavy on his hands; he could not sleep. He achieved a little publicity by claiming that he had not slept for four years and four months because of ill health. "Nunquam dormio," he wrote to the papers, and solicitous readers wrote to inquire about his health. The servant who lit his fire in the morning claimed that he caught him sleeping. Gourlay said he was merely warming his head under the covers. Hotel residents said he napped in the lobby, but he claimed he was just resting profoundly.[15] It seemed a low level of occupation for the man who had been a gadfly. He was not the only aging bore. Robert Owen was becoming tiresome and repetitious, and the great Babbage was raging from his window at organ grinders.

He wrote in his Journal: "I flatter myself something *useful* may be

---

[12]*The Banished Briton*, #7, p. 68.

[13]Gourlay Papers, OA, 1842.

[14]The Combinations Act would be employed as late as 1872 when George Brown, editor of the Toronto *Globe*, had twenty-four printers arrested on a charge of seditious conspiracy for agitating for a nine-hour day. Doris French, *Faith, Sweat, and Politics*, McClelland, 1962.

[15]*The Banished Briton*, #38, p. 511.

extracted—after a lifetime of seeming uselessness—certainly, no Individual . . . ever had hopes more completely blasted & projects so utterly discomfited—yes, it may be the will of God, after all, to make use of the stone despised—for a corner—and, at last, if no seen results—at last there will be at least an opportunity for the best of daughters—still—still—to regard their father—as one who can be remade—forgiven, & received with them into the other and better world."[16]

He returned to Canada and asked his friends to allow him to appear before the Bar of the House. The only thing that could be changed was the amount of the pension. "Tiger" Dunlop was exasperated: ". . . I don't know what you want. I got you to draw the prayer of your own petition, and, by a unanimous vote of the House, got you all you wanted, and more. You put douce Davy Thorburn in bodily fear, by showing your crooked stick on the floor; he thinking you would make a Lord Brougham of him. When I find out what you want, I'll try and get it; but it is the d ———— l to do with Fife folk, who are all daft and should be confined in an exclusive lunatic asylum, at Cupar,—keepers being had from the neighbouring Counties of Clackmanan, Kinross, Perthshire, and the Lothians."[17]

In September 1845, he was back in Montreal to importune the latest governor, Sir Charles Metcalfe, for Bagot had been recalled. Metcalfe invited him to dinner in the course of which Gourlay solicited his patronage for a second statistical account. Metcalfe governed for the last nine months of his tenure without his ministers, reform or otherwise. Responsible government would have to evolve without help from either Gourlay or his friend Tiger Dunlop. Neither believed in it, although it would eventually solve the troublesome problem of colonial autonomy within the empire.

Again erysipelas plagued Gourlay. In October he decided to "enjoy a horn" with Dunlop, the giant Highlander whose legs had calves as big as casks. Setting off on foot from Galt, where William Dickson now resided prosperous and respected, his swollen leg caused him to proceed in "a genteel vehicle". He stayed a week at Gairbraid, high over the mouth of the Maitland River where it emptied into Georgian Bay, and then returned to Queenston by steamboat.[18] In the spring he asked again to be heard by the legislature. The only equal to Gourlay's persistence was the British monarchy that had not renounced its claim to the French throne and the quartering of the fleur de lys on the royal coat of arms till 1801. In May 1846, Gourlay returned to Scotland, the land where he wished his bones to lie.

16Boston, Sept. 30, 1843, Gourlay Papers, OA.
17*The Banished Briton*, #36, p. 492.
18*Ibid.*, p. 491; #39, p. 24.

25

Man is a Recording Animal

England had seen many changes since Gourlay left it in 1833. The term "re-dundant poor" was replaced by the "working class".[1] The inadequacies of the Reform Bill of 1832 and the Poor Law of 1834 gave rise to the Chartist movement. Its methods were the same as Gourlay's—peaceful petitioning and the calling of a convention to waken parliament to its duty. The Chartists adopted a six point charter and amassed two petitions, one in 1839 with 1,280,000 signatures and one in 1842 with 3,315,752. Had they adopted Gourlay's orderly method of petitioning by parishes, they would not have been accused of falsifying signatures. In 1839 they called an elected convention that met first in London and then in Birmingham. The Chartist movement failed because the Commons would not accept its petition, and it had no alternative plan in case of failure. Robert Gourlay, the ostensible failure, had provided continuing committees to carry on the work of his Convention.

The Corn Laws that Gourlay had fought when he was in Wiltshire still remained. In 1845 came the rain that nourished the blight that blackened the potato plants before the tubers formed. Whig Lord John Russell said: "The state of Ireland for the next few months must be one of great suffering." He refused to supply new seed. Whig Lord Brougham said: "Undoubtedly it was the landlord's right to do as he pleased, and if he abstained [from evicting tenants and pulling down huts], he conferred a favour and was doing an act of kindness."[2] The Irish peasants poured out of their stricken land to Canada, dying of fever on the way or in fever sheds in Montreal. There was no guilt for sins of omission on the conscience of the man who for years had advocated planned emigration and reviled the government for forcing the English and Irish peasant to a potato subsistence. A fate so horrible that no one could have envisioned it had befallen the Irish peasant. How was the English peasant faring? As long

---

1*The Banished Briton*, #7, p. 69.
2Cecil Woodham-Smith, *The Great Hunger*, pp. 65, 317, 72.

after as 1853, the Wiltshire papers report, bands of 100 to 150 quiet but determined men were going from farm to farm asking for wages higher than eight shillings so they could save their families from degradation. The only thing that could be said for advancing times was that they were not treated like the Tolpuddle martyrs of 1834.

Gourlay still thought he would like to enter parliament. Though he was sixty-eight, he contested the election that ended Wemyss' long tenure in Fife. All he asked of the electors now were "a bed, an umbrella and a bannock".[3] At another election, wrote Mathew Conolly in *Biographical Dictionary of Eminent Men in Fife*, 1866, Gourlay nominated himself as a candidate though he was told it was an impossibility. Conolly relished Gourlay's eccentricities, but he described as well the effectiveness of his clear, well-controlled voice, his vehement but graceful actions, and his long gesticulating arms that he made almost to speak. His visits to Cupar were always enjoyed. On one occasion he gathered a group around him, and for a platform he mounted the steps of the Kingdom of Fife coach in front of the Horse and Dog at the foot of the Long Wynd. Some boys, wanting to have a little fun with greybeard, began to pull the coach about. Entirely uncommoded, Gourlay continued to speak and pass out pamphlets, pausing in the midst of his proposals for free trade to caution the boys to be careful to do no mischief with their coach pulling.

Daily happiness was now his in spite of breaks in his family circle. His son Oliver had died on November 19, 1843, on board ship on a trading journey from Port Sydney, Australia, to Manila and China. His father was proud to learn from the captain that on one occasion his son had saved several people from drowning and, after being sworn in at Port Philip as a special constable to go into the bush after Bush Rangers, had so distinguished himself that a public dinner had been accorded him.[4] His youngest daughter, Catherine, who was happily married to Mr. Alexander Duncan of Parkhill, Arbroath, had died in childbirth, 1844. Little Alexander Robert became his grandfather's delight. Helen, who had been in France as a governess, returned to keep house for her father at Sunnyside Cottage, Montrose, not far from Arbroath. Debts of £2,437 that had tripled themselves with interest of £5,743 were paid off at seven shillings nine pence per pound in June 1847 and February 1849.[5]

In 1849 Gourlay advertised for fifty men to go to Canada to settle his land in Dereham Township. He published details in *Emigration and Settlement on Wild Land* for the benefit of enquirers but nothing came of it. He visited friends in Edinburgh including Dr. Chalmers. He was still pressing for improvements to The Mound in Edinburgh, and was now

---

3*To the County Electors of Fife*, 1846, p. 11.
4Boston, Dec. 25, 1844, Gourlay Papers, OA.
5Gourlay Papers, OA.

advocating free wheat from Canada and dutied wheat from the United States. It was Peel's proposal of free trade except for one shilling a bushel that was accepted this time. After he addressed a public meeting in Edinburgh in 1852, he was knocked down by a carriage and suffered a broken thigh that left him permanently lame.

In the same year, his seventy-fifth, he began to finish his autobiography, parts of which were already printed in previous publications. A tentative index indicated a concluding chapter: Man is a Recording Animal—Your Soul—My Soul—The Soul.[6] In script that signified that his hand had lost its strength, he wrote fourteen pages recounting boyhood events, and then discontinued it, for he had decided to make a last visit to friends in Canada while yet there was time. Accompanied by Helen, he landed in Philadelphia on August 29, 1856. On September 23 he was in Canada where he announced his arrival in the papers with his customary flourish: "Call to Canada West". He and Helen toured and admired the Rideau Canal system, visited the Hon. John Hamilton in Kingston, took the train to Ingersoll with no criticism of the service. While his daughter visited in Virginia before returning to Scotland, Gourlay enlarged his house at Dereham to two storeys and built a carriage house behind it with rooms over it for a servant. He engaged a young Irish woman named Mary Reenan as his housekeeper and John Smith of nearby Mount Elgin as his man of business. He made friends with the editor of the *Ingersoll Chronicle* and reprinted *Chronicles of Canada*. He left the numbers of *The Banished Briton* with the printer who evidently bound them in the unchronological order that has exasperated readers ever since.

He was still convinced that the exoneration granted him in 1842 had been insufficient. In wavering script he asked to be heard at the Bar of the House. Indulgent laughter ran through the House at the request but on May 21, 1858, permission was granted, 41-38. Alexander Mackenzie and George Brown voted for the hearing, and John A. Macdonald and John Sandfield Macdonald voted against it. On the day appointed, Gourlay was ill, but on June 30 he was able to state his case. He said that the 1842 pardon implied guilt on his part; he still wished his treatment declared unjust from the beginning. As usual he printed the proceedings in a pamphlet entitled *Mr. Gourlay's Case before the Legislature*, July 1, 1858.

Eighteen fifty-eight was the year of the election in which the loudest slogan was "Rep. by Pop.", and confederation was already an issue. Among the numerous candidates for a seat representing Oxford County was Robert Fleming Gourlay, aged eighty. James MacIntyre, one of the local poetasters known later as the "Cheese Poet", recorded the result:

6*Ibid.* This printed one-page index was found during a second trip to Scotland when over half this biography was in first draft.

Robert Fleming Gourlay

There came to Oxford Robert Gourlay
In his old age his health was poorly;
He was a relic of the past,
In his dotage was sinking fast;
Yet he was erect and tall
Like noble ruined castle wall.
In early times they did him impeach
For demanding right of speech,
Now Oxford he wished to represent
In Canadian parliament;
Him the riding did not honour,
But elected Doctor Connor.[7]

Gourlay received one vote in this election of many candidates; it was not his own. One of the other candidates received none.

He was still capable of surprising people. This same year he married Mary Reenan, his twenty-eight-year-old housekeeper, who had been serving him commendably. Attracted by his courtly manner and his carriage house, Mary Reenan probably thought that the old man had money. As soon as she married and found that even the land from which he fought off squatters as fiercely as if it were his own was still in another's name, she changed to a shrew. On more than one occasion "old man Gourlay" fled for peace to John Smith at nearby Mount Elgin. It was soon too much for him and leaving the land, which like Circe had enticed him to Canada in the first place, he returned to Edinburgh to live with his daughters.[8]

Once again he called a meeting in Cupar. With two canes to hold him erect, he walked to the platform of the Tontine to address a small audience on emigration. When he called for questions, a town resident began to ridicule his ideas on political economy. Gourlay asked him to repeat his question as he was hard of hearing. He pretended not to understand him and made some observations, totally different from the subject, in such a way that the audience ended up laughing at the questioner. Gourlay was game to the end.[9]

Still his autobiography was not finished. Before he left for Canada in 1856 he had boxed his books and papers and sent them to Cupar by train with instructions to the baggage-master to hold them till he arrived. Instead of returning to Cupar he had set out for Canada. When he called

[7]W. A. Deacon, *The Four Jameses* (Graphic, Toronto, 1927), p. 57.

[8]*Ingersoll Chronicle*, 1858 file, in possession of Stanley J. Smith, Ingersoll; Farm Book, Parkhill, Arbroath; Land Records Office, Woodstock, Ontario; "Robert (Fleming) Gourlay—Reminiscences of his last days in Canada", Mrs. Sidney Farmer, OH, 1917.

[9]M. Conolly, *Biographical Dictionary of Eminent Men of Fife*.

at the station for the boxes that he had despatched four years before, they were gone, thrown out as unclaimed goods.[10] Gourlay's rage was Lear-like. It seemed an impossibility to gather his record again. Nevertheless on December 31, 1862, in his eighty-fifth year, he took up his pen. "Man is a recording animal", he had written in 1829 on the cover of *Chronicles of Canada: Being the Record of Robert Gourlay, Esquire,* and he was living up to his words. He described the time when he was still in petticoats and the press-gang had frightened him when they came to his grandfather's farm at Frierton to seize the young men. He recounted remarkable stories related to him by a little Miss Gourlay from Kincraig whom he had met in Edinburgh before leaving for his last trip to Canada. She told him about her grandfather, William Gourlay, the twenty-first laird of Kincraig, the promontory that Ingelramus de Gourlay had been granted for watching for invaders at the entrance of the Firth of Forth. William was called Strong Gourlay. He had pulled his horse out of a bog as if were a puddock (frog); had fought an oppressive laird till he capitulated crying, "Ye're either the De'il or Gourlay of Kincraig"; had linked together with a twisted iron spit two Englishmen who had preëmpted the fowl he had ordered for dinner in an inn; had lifted the lintel of his new house into place when two masons had gone to recoup their strength with dinner.[11]

Robert wrote of his martyred ancestor Norman Gourlay, though he did not include in the record John Knox's summation of Norman Gourlay's characteristics that had come down to himself inescapably through the centuries: "Others abjured and publictlie brynt thare byllis; otheris compeared not, and tharefoir war exiled. Butt in judgement war produced two, to wit, David Stratoun, a gentilman, and Maister Normound Gowrlay, a man of reassonable eruditioun, of whom we mon schortlye speak. In Maister Normound appeared knawledge, albeit joyned with weakness."[12] Normound was murdered and burned; Robert was banished. Neither held their peace in order to preserve themselves.

Early in 1863 Gourlay suffered a stroke that left part of his face paralyzed, but still he would not acknowledge that he was finished with life. With one last shake of his crooked stick in the face of fate, he dressed for the photographer so that he could send another picture to his friends in Canada. The well-kept hair and beard wore now the unkempt quality of conquering age and the face was marred by the stroke. Soon after, there came to pass the only event in Robert Gourlay's life that he could not record. On August 1, 1863, his body struggled to expel the spirit it no longer had the strength to contain. At 20 Howe Street, Edinburgh, just south of Thistle Street, Robert Fleming Gourlay drew his last breath

---

10*Ibid.*
11 1852 ms., Gourlay Papers, OA.
12 J. Knox, *History of the Reformation,* I, p. 58.

behind the lace curtains that covered the wide window. The old Presbyterian roustabout had departed for "the eternity of bliss beyond the
grave".[13]

They laid Robert Gourlay to rest beside his parents in the box tomb
in Ceres churchyard close to the mausoleum that housed the remains of
the Crawfords, senior aristocrats of Fife. He had prepared the inscription
himself, tracing his lineage back to Ingelramus de Gourlaye. The wind, the
rain and the lichens have blurred the words on the soft Fife sandstone, but
the written record remains.

His will showed him possessor of less than £200 in money and two £10
shares in the Edinburgh Workmen's Improvement Company. In a note at
the bottom, daughter Jane wrote finis to her father's obsession: "The accused had a claim on the Canadian government for alleged injuries sustained by him many years ago. No value can be put on it." Nevertheless she
cannily applied for the pension that her father had spurned with his
customary grandiloquence. The Honourable George Brown presented the
daughters' claim to the Canadian government and in 1859 an order-in-
council provided for payment retroactive to 1842 of the $200 pension that
Gourlay considered was beneath him. On April 13, 1866, provision was
made for payment of $4,361.10 to the Misses Gourlay.

He bequeathed the land in Canada to his grandson, now studying for
law. The title was still not clearly in the Gourlay name. Shortly after her
elderly husband returned to Scotland, Mary Reenan Gourlay disappeared.
Her brother thought she might have committed suicide. A man called
Riddle cut wood illegally on the property. He found Mary in Toronto and
swindled her out of the land, registering it in the name of an infant son
as he had lawsuits pending against himself. In 1871 the ownership tangle
was unsnarled and the land registered in the names of Jean and Helen
Gourlay, who turned it over to Alexander Duncan. Duncan's interest in it
was sufficient to bring him to Canada, but when he arrived in Toronto he
did not even go to look at his land in Dereham. Lockhart Gordon, a law
school friend whom he had advised to go to Canada to practise, persuaded
him to buy land in the Haliburton area for he considered it was the coming
vacation spot in Ontario. Alex Duncan sold the rich Oxford County land to
two brothers, J. S. and C. J. Banbury. The house that Gourlay had built
burned down when rented to a family of negroes who threw hot ashes on
the wooden verandah floor.[14] It was a long contentious time since the land
had been acquired by Robert Hamilton in 1800.

It was a long time too since Oliver Gourlay had said, "Robert will hurt

---

[13] *Appeal*, p. lxii.
[14] Conversation with Mr. James Banbury, Woodstock, whose father and uncle bought
Lot 3, Concession 4, from Alexander Duncan in 1878. They helped extinguish the fire,
and also assisted Gourlay in evicting squatters from the back section of his property.

himself but do good to others." It was Robert Gourlay's fate to be born with the compulsion to goad till wrongs were remedied. Social and political movements tend to stagnate till they are given impetus and publicity by the gadflies who come at their beginning. Contemporaries usually give a gadfly so much more scorn than praise that he is regarded as a failure. Though no specific success can be credited to the man whose life has been delineated here, no one may ridicule the fearlessness he displayed in the defence of great causes. Because of that, the name of Robert Gourlay will never be consigned to the limbo of the little.

PUBLICATIONS OF ROBERT (FLEMING) GOURLAY

"Inquiry into the State of the Cottagers of Lincoln and Rutland", *Annals of Agriculture*, Vol. 37, pp. 514-550.

Letter to the Earl of Kellie concerning the farmers' income tax, with a hint on the principle of representation, Ballintine & Law, London, 1808, pp. 63.

A Specific Plan for Organizing the People and for obtaining Reform Independent of Parliament—to The People of Fife . . . of Britain!, printed for the author, Ballintine and Law, London, published by J. M. Richardson, 23, Cornhill, London, 1809, pp. 179.

Apology for Scotch Farmers in England with a Case in Tythes, submitted to the consideration of members of the Bath and West of England, and Wiltshire Agricultural Societies, Highley and Son, London, 1813, pp. 23, app. vii.

"Address to Farmers of the Hill Country of Wilts, Hants, and Dorset", Salisbury Journal, Nov. 21, 1814.

"Institution for the Benefit and Protection of the Farming Interest", Jan., 1815.

The Right to church property secured and commutation of tithes vindicated, in a letter to the Rev. William Coxe, Highley & son, London, Feb. 24, 1815, p. 41.

An Address to the Special Jurymen of Wiltshire: with a report of two issue trials at Salisbury with copies of documents in Ms., Henry Gye, Bath, Mar. 12, 1816, pp. lx and 200.

Liberty of the Press asserted in an appeal to the inhabitants of Wilts and a letter on the corn laws, Meyler & son, Bath, 1815, pp. 16.

Tyranny of Poor Laws exemplified, Gye & son, Bath, March 13, 1815, pp. 15.

To the labouring poor of Wily Parish, H. Gye, Bath, 1816, pp. 21.

The village system; being a scheme for the gradual abolition of pauperism, and immediate employment and provisioning of the people, H. Gye, Bath, 1817, pp. 40.

The petition for the benefit of the labouring poor, presented and not presented by Sir Francis Burdett, discarded by Lord Cochrane, and spurned by Lord Folkestone, now laid before parliament, with occasional correspondence and remarks on the subject of the poor laws, parliamentary reform, etc., Henry Gye, Bath, March, 1817, pp. 42.

(First Address) "To the Resident Land-Owners of Upper Canada", *Upper Canada Gazette*, York, October 30, 1817.

268

(Second Address) "To the Resident Land Owners of Upper Canada", *Niagara Spectator*, Feb. 5 and 12, 1818.

(Third Address) "To the Resident Land-Owners of Upper Canada", *Niagara Spectator*, April, 1818.

Principles and Proceedings of the Inhabitants of the District of Niagara for addressing his Royal Highness the Prince Regent respecting claims of Sufferers in War, Lands to Militiamen, and the general benefit of Upper Canada, printed at the *Niagara Spectator* office, April, 1818.

Letters to the resident landowners of Upper Canada, by RG & replies to Mr. G., Kingston, April, 1818, pp. 167.

Narrative of a Journey through the Midland, Johnstown, Eastern and Ottawa Districts, to publish Principles and Proceedings of the Inhabitants of the District of Niagara, printed at *Spectator* office, Niagara, August, 1818.

Address to the Jury at Kingston assizes, 1818, in the case of the King vs. Robert Gourlay for libel, with a report of the trial, etc., printed at the *Gazette* office, Kingston, August, 1818, pp. 24.

(Fourth Address) "To the Resident Landowners of Upper Canada", Niagara Jail, May 20, 1819, published in *Niagara Spectator*, May 27, 1819.

Statistical Account of Upper Canada compiled with a view to a grand system of emigration in connexion with a reform of the poor laws, 2 vols., 1822, Vol. I, intr. pp. div, 625; vol. 2, intr. xxii, 704, app. cxxix, index. Reprinted 1966. Johnson Reprint Corporation, New York.

An Appeal to the Common Sense, Mind, and Manhood of the British Nation, London, printed for the author, 1826, pp. 196.

Plans for the Improvement of Edinburgh, Edinburgh, 1829, pp. 38.

The Record—dedicated to the labouring classes of Fife, Edinburgh, pp. 32.

Chronicles of Canada: Being a Record of Robert Gourlay, esq., now Robert Fleming Gourlay, "the Banished Briton" . . . No. 1, Concerning the Convention and gagging law, 1818, Mr. Gourlay's arrest and trial, etc., St. Catharines, printed at the Journal office, 1842, pp. 40.

Record of my Private Affairs, 1832, not extant.

Mr. Gourlay's Addresses etc. to the electors of Fife, G. S. Tullis, Cupar, 1847, pp. 12.

Proposals for drawing up and publishing a statistical account of Ohio: under the direction of its legislature, Cleveland, Ohio, 1836.

The Banished Briton, appellant and mediator, Cleveland, Ohio, 1836, pp. 16.

"Plan for Kingston", *Kingston Gazette*, August 21, 1841.

"Resolved that a humble address be presented to His Excellency the Governor-General communicating a copy of the report of a select committee of this house on the case of Robert Gourlay, Sept. 16, 1841.

Correspondence with Mr. Hume, 1829-32, (1852?) pp. 3.

The Banished Briton and Neptunian: being a Record of the Life, Writings, Principles, and Projects of Robert Gourlay, Esquire, now Robert Fleming Gourlay, S. N. Dickinson, Boston, 1843, pp. 512 plus #39, pp. 24.

Plans for beautifying New York and for enlarging and improving the City of Boston—being studies to illustrate the science of city building, Boston, 1844.

Mr. Gourlay's Addresses, etc., to electors of Fife, Cupar, G. S. Tullis, Printer, 1847(?) 12pp.

Emigration and settlement on wild land, Cupar-Fife, 1849, pp. 20.

The Mound Improvement, with a plan and elevations; also an appendix, containing correspondence with Dr. Chalmers, the City authorities, etc., etc., on the same subject, Adam and Charles Black, Edinburgh, 1850, pp. 40.

The Mound Murder, and the Mound's Appeal, correspondence with the Lord Provost, etc., Edinburgh, 1852, pp. 19.

Canada and Corn-Laws; or No Corn Laws, no Canada, James Wood, Edinburgh, 1852, pp. 12.

The Best Site for Trinity College Church: with ten years' discussion on the subject and seven lithographs, Edinburgh, 1855.

Manual of Worship, Edinburgh, 1856, pp. 25.

Mr. Gourlay's Case before the Legislature, with his speech delivered on Wednesday, July 1, 1859, Globe office, Toronto, pp. 29.

For bibliography of works of other authors consulted see page notes.

Abbreviations used: Jls. Leg. Ass.—Journals of the Legislative Assembly; TPL—Toronto Public Library; OA—Ontario Archives; PAC—Public Archives of Canada, Ottawa; OH—*Ontario History* magazine; CHR—*Canadian Historical Review*.

Index

PLAN
OF
KINGSTON HARBOUR, &c.

Scale.

MAP
OF
UPPER CANADA

Engraved for Statistical Account.

N.B. The Circles & Lines on the unsurveyed tracts are to illustrate
the Compiler's Plan of laying out these tracts. The large circle represents the size of a New Capital city...the smaller,
County-towns...the smallest, Towns. Double dotted lines.
grand central high ways...single dotted lines County
roads & lines, Township roads.

Sketch of the Practicable courses of the
GRAND COMMERCIAL CANAL or St. LAWRENCE,
with its Junctions.

Scale

Published by Longn.